A Refutation of Evolution

The Intelligent Design of Man

by

Edmund T. Dombrowski, MD

Published by MWP Media
30092 Alicia Parkway, Laguna Niguel, CA 92677

Cover designed by Greg and Deborah Pry

ISBN
Printed in the United States of America

Dedication

This book is dedicated to my family: my dearest wife, Ellie, my sons Edmund Theodore III, Mark Stephen, and my daughters, Kathleen Ann, Linda Lee, and Deborah Lucile.

I wish to also thank those who have contributed to knowledge for knowledge sake and that which was correct in the past as well as that which is correct now and will forever be correct on the subject of Intelligent Design.

—Author: Edmund T. Dombrowski

Acknowledgments

I would like to express my appreciation to a number of people who helped in the preparation of this book:

Jim Adams who made the slides that are noted throughout the book. His e-mail address is sccjim@yahoo.com.

Greg and Deborah Pry (my youngest daughter) who made the cover of the book. Deborah also made the bird diagram, showing the way the bird lung works and many other drawings.

My daughters, Kathleen Ann and Linda Lee, helped in the initial proof reading.

My son, Mark Steven, helped keep my computer going.

My wife, Eleanor Lucille, helped in many, many ways.

My special thanks to you all!

Table of Contents

Preface

In this, the early 21st-century, it is a sad and disturbing commentary on the science of biology that so many teachers, at all levels of education, continue to teach Darwinism as though macroevolution were a proven fact; and as though macroevolution were the only explanation for life's diversity – which it certainly is not.

Ignoring fact is foolish, particularly scientific fact. It is strange that those who indulge in these unsubstantiated claims appear to enjoy self-delusion. Here is an interesting thought about evolution:

The probability of the complex birth of an infant, concerning integration of modalities that would have to have been put in place by chance origin, is determined by multiplying the number of modalities involved. When this is done, the negative probability is one chance in 10,000 billion billion billion billion that the integration might have been put in place by chance, if then. This doesn't take into account the formation of the maternal or infant bodies, or the production of the enzymatic or structural proteins required for these modalities. Probability states that this system was integrated by *intelligent design and irreducible complexity*. You have to ask, what rational person would base their philosophy of life on such a negative probability, let alone bet their careers.

There is no reason to deceive ourselves as to who we are, or even why we exist at all. As rational human beings, it is important that we devote time to addressing the most important judgment we will ever make. Do we exist by chance or by design? There are many examples of empirical evidence that refute macroevolution from the standpoint of probability.

It is not that evolutionists are necessarily disingenuous, rather, the problem lies with the small bubbles of understanding in which single entity scientists live and work. They just don't know the areas of science that preclude macroevolution. In addition, the field of biology, improperly employed, is circular thinking. This sort of reasoning is imprecise and indiscriminate that results in improper nomenclature.

This circumstance alone, with a humanist philosophy and a desire to place the origin of life at a zero, inorganic starting point, leads to a highly slanted bias toward macroevolution. Then too, there is a propensity among many teachers, journalists and politicians to give way to academic authority as a substitute for personal study and logical evaluation of the evidence available. In their blind adherence to evolutionist dogma, they mistakenly inculcate and perpetuate Darwinism.

This is a circumstance which cannot be allowed to continue, when a design hypothesis is a far better, evidence-based explanation for life's biology than is the gap-ridden Darwinian macroevolution hypothesis.

The design hypothesis is straightforward, easy to understand, conforms to all observed facts, and obeys Ocham's razor (meaning, entities must not be multiplied beyond necessity and the conclusion thereof, that the simplest explanation or strategy tends to be the best one).

The Earth on which we live, the solar system and our position in it, the sequence of events that fashioned the Earth's interior, and the Earth's mantle, our biosphere, all life forms, and our bodies are the products of specific design; fashioned into a system of ecostasis on the basis of properly integrating and utilizing the dynamic interaction of the basic laws of nature, and this is in conformity with the laws of probability as contrasted to spontaneous self–organization from inorganic material and subsequent chance, non-goal-directed sequential random change.

FOR EXAMPLE, TO DESIGN an organ as complex as the human eye requires multiple paths that must be independently and simultaneously followed in such a way as to reach their required goals at appropriate corresponding times, while integrating the improbability of their accomplishments, and all at the same time.

That is the function of the human eye that requires a modification of the skull with an eye socket into which the eye can be placed and protected. At the same time, muscles have to be developed within the eye socket in such a way as to produce complete motor control in all planes. This requires six muscles for each eye; all of which have to be integrated and supplied with appropriate motor and sensory nerves.

It is impossible to believe that this integration could have been carried out on a stepwise (marked by a gradual progression) chance basis, independently in each system, and to have it just happen a step at a time in each system, and end up with functional, movable eyes with both usable binocular vision and usable independent lateral vision during the transition period from one to the other.

In the case of the superior oblique muscle in the eye, which produces downward and outward gaze, there would have to have been a little pulley in the eye socket, over which the muscle could change the direction of its pull, while still maintaining suitable locations for its origin and insertion.

There is yet another insurmountable problem with the concept of chance stepwise origin of the eye. The medial, lateral, superior, and inferior rectus muscles exert their pull only backward into the back of the eye socket. This means that if only the rectus muscles were in play, the eyeball would have been jammed against the posterior wall of the eye socket at all times, making it difficult for the eyeball to move, while at the same time jamming the optic nerve against the edges of its foramen. The pull of the two oblique muscles, the superior oblique and the inferior oblique, counteract the pull of the rectus muscles and keep the eyeball centered in the eye socket.

What would have happened to binocular vision if all six ocular muscles had not formed at the same time, in both eyes, and with integrated nerve supplies so as to allow ocular fusion of the visual panorama displayed on the right and left optic cortices? The answer is that any organism without a full complement of muscles lacking a full set of nerves, void of an integrated display on the optic lobes of the brain, could not have survived. Is this not a case of irreducible complexity? Irreducible complexity, or IC, is an argument by proponents of intelligent design that certain biological systems are too complex to have evolved from simpler, or less complete predecessors, through natural selection, acting upon a series of advantageous naturally occurring chance mutations.

Who is ready to step-forward and scientifically chart the method and the sequence by which statistically-random changes/mutations might occur at various locations on various genes along a chromosome, and at various times such that the resulting mutations would be correlated with the required changes and capable of producing the integrated anatomical and functional modifications required to convert independent lateral gaze into unified binocular gaze as outlined above?

It is doubtful there will be any volunteers ready to explain the above sequence of events, because it is impossible to explain the impossible. First, keep in mind that man can accurately observe his surroundings and formulate laws to describe them (this is called the observable axiom). Second is that every event that has happened or is happening can be explained by math, chemistry, physics, or biology (the naturalistic axiom). Science requires both axioms to function properly.

DESIGN OF THE SPECIES IS PRESENTED as a challenge to macroevolution's sophistry. It is written here to expose the many mistaken tenets of Darwinism and to promote a more precise understanding of life in terms of the factual evidence at hand – not blind adherence to the dogma of a bygone era.

Part I

Intelligent Design vs. Macroevolution

Chapter One
Basic Tenets of Darwinian Concept of Descent

In the case for natural selection and macroevolution refutation, let's be sure we all understand the basic concepts of Darwinism using a presentation of the concept as it appears in a college-level textbook of biology published in 2003. The textbook presents the essentials of Darwinism and neo-Darwinism descent-with-modification as follows:

A population is a group of individuals of the same species living in the same place at the same time. It is the smallest unit that can evolve. Darwin understood that it is populations that evolve. He saw that natural selection, in favoring some heritable traits over others, could change populations over successive generations. What eluded him was the genetic basis of population change. Without this basis, he could not explain the cause of variation among the individuals making up a population, nor could he account for the perpetuation of parents' traits in their offspring.

An important turning point for evolutionary theory was the birth in the 1920s of population genetics, the science of genetic change in populations. A theory of evolution that includes genetics was developed in the early 1940s. Known as the modern synthesis, also called neo-Darwinism, this theory focuses on populations as the units of evolution and includes most of Darwin's ideas.

Most importantly, it melds genetics with the theory of natural selection. Central to the modern synthesis is the relationship between populations and species. But for now, let's limit our discussion to organisms that reproduce sexually. In this context, we may define a species as a group of populations whose individuals have the potential to interbreed and produce fertile offspring. Each species is distributed over a geographic range, where individuals usually concentrate in localized populations.

One of the more popular advocates of macroevolution is Dawkins, who studied at Oxford and graduated in 1962, who is a Ph.D., and became an assistant professor of zoology at the University of California, Berkeley, from 1967-1969. He later returned to Oxford to become a lecturer in zoology and fellow of the New College in 1970. He became the first holder of the Charles Simonyi Chair of Public Understanding of Science, which he continues to hold at the time of this writing. He has written several books from the standpoint of an atheist that have received some praise among evolutionists. He refutes the concept of intelligent design in favor of macroevolution.

Without realizing it, Professor Dawkins characterized and summarized the argument between design and evolution in favor of design, apparently without recognizing his mistake. Professor Dawkins' choice of a simile can be compared to Berrera's blunder. Tim Berrera tried to compare the fossil changes noted at various depths of the Cambrian layer of geologic strata to the metamorphosis of a 1953 Corvette into a 1978 model. In citing a series of industrial design changes as the simile for his postulates, he only succeeded in destroying his own argument by illustrating the effect of design.

Continuing in chapter three of Dawkins' book, he writes: "…living things are too improbable and too beautifully 'designed' to have come into existence by chance. How then did they come into existence? The answer, Darwin's answer, is by gradual step-by-step transformations from simple beginnings, from primordial entities sufficiently simple to come into existence by chance. Each successive change in the gradual evolutionary process was simple enough, *relative to its predecessor*, to have arisen by chance. But the whole sequence of cumulative steps constitutes anything but a chance process when you consider the complexity of the final end-product relative to the original starting point. The cumulative process is directed by nonrandom survival."

In addition, Professor Dawkins forgets that nonrandom survival is ultimately a consequence of the uncontrollable, random molecular "mutations" that occur within a population's genome. Therefore, nonrandom survival is, itself, the result of chance.

Secondly, Professor Dawkins recognizes that each "successive change was simple enough to have arisen by chance." However, he then states "but the whole sequence of cumulative steps constitutes anything but a chance process." This is pure speculation and an obvious non-sequitur. Dr. Dawkins offers no explanation for the method by which chance steps were converted into "anything but a random process." No one, not even a professor, is allowed to make such statements without offering a description of the mechanism by which chance was converted into a nonrandom process.

Later, in the same chapter, Professor Dawkins embellished his opinion as follows: "Darwinism explains living organization in terms of chance 'single-step selection' alone. This belief that Darwinian evolution is 'random' is not merely false, it is the exact opposite of reality. Chance is a minor ingredient in the Darwinian recipe, but the most important ingredient is the cumulative selection which is quintessentially nonrandom. Cumulative natural selection is blind to the future and has no long-term goal."

The above comments represent a disturbing example of imprecise teleological thinking. In effect, Professor Dawkins infers that the end results of chance have turned out the way they have, because that's the way things have turned out. Such thinking is not in keeping with the thought processes of science. By what right, and by what evidence, does Dr. Dawkins proclaim that “selection (is) quintessentially nonrandom”?

It is difficult to understand how a person of Professor Dawkins’ caliber could have written this litany of non-sequiturs.

Chapter Two
Evolution, a Failed Theory

The most important question any human can ever ask or answer is whether he or she exists by accident or by design. If we exist by accident, then the philosophy of the humanists and secular progressives makes sense. But, if all the ecosystems of the Earth are designed for the purpose of maintaining the species in the world, and if homeostasis (a point of balance or internal equilibrium) is the cornerstone of physiology, then each one of us should carefully consider the implications of living on a carefully designed planet in elegantly designed bodies.

A Darwinian macroevolutionist is one who believes in the theory of evolution. More especially, one who tries to point out where creationist views conflict with the accepted scientific view. This term, macroevolution, carries an implication of vocal opposition to the creationist viewpoint.

This treatise is particularly dedicated to informing the inquisitive reader about the overwhelming evidence for design that exists in biologic systems – information that is known, in-depth, and with certainty, to all applied scientists; those practical scientists who are guided by the demands of evidence-based reasoning, and whose opinions are molded by reality and legal liability. It is written for three reasons:

The first is because of the amazingly compelling evidence of precise design inherent in the normal form and function of the human body; evidence which demands an overt presence in the minds of all humans who appreciate a comprehensive and reliable understanding of the world in which we live. In the presence of such an understanding, the spurious concepts of "natural selection" and the "plausible excuse for evolution's gaps" can be recognized as the false precepts they are.

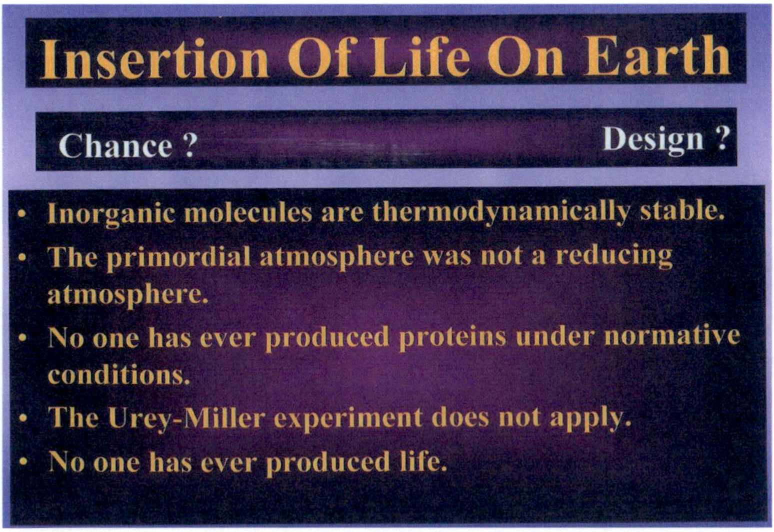

Once recognized as untenable explanations for the process of macroevolution, these proponents lose their unsubstantiated ability to mislead developing minds into believing that plants and animals can be transmuted from one species of a lesser complexity into another of greater complexity.

It is very clear in science that organisms of increasing complexity demonstrate increasing amounts of nuclear chromatin. However, there is no mechanism by which a cell can produce increased amounts of chromatin, that is DNA. Why? This is so because all cells reproduce by duplicating already existing DNA molecules, molecule for molecule. Although it is true that there can be minor mutations, still there cannot be any overall increase in genetic lines of code because no additional chromatin is ever produced.

Moreover, evolution from one species into another has never been anything more than an unproven speculation; it is something that has never been demonstrated, and therefore a spurious unproven postulate without factual support.

A priori (used in philosophy to distinguish two types of knowledge), the concepts of natural selection and genetic drift are illogical when considered as mechanisms capable of producing an increase in genetic complexity.

How can the selecting out and removal of portions of an organism's genome (existing lines of code) produce increased complexity, when some of genome's previously possible permutations and combinations have been removed? After selection, there are fewer possible permutations and combinations to work with than there were before.

This is something evolutionists have never explained. Rather, they postulate a highly unlikely scenario, suggesting that existing genes are reshuffled to become super genes, or switch genes when randomly mutated into new locations within a DNA molecule. This is a possibility only as regards minor variations, such as a change in color or texture. However this type of "mutation" has never been demonstrated to produce a new, advantageous functional modification in an existing species.

In the final analysis, the effect of "natural selection" would be to weed out the extremes of potential variations inherent within a genome's possible phenotypes, and thereby maintaining a genome as an identifiable species. In other words, natural selection, in reality, maintains the stability of the species. The effect of genetic drift is to produce nothing more than a limited population of variant phenotypes skewed to one side or the other of a genome's expected norm. Genetic drift does not produce an increase in available genetic information.

Genetic drift simply produces variance from the general norm that is intra-species variation.

For evolutionists to continue promulgating the bigotry of a humanistic finite zero starting point, as they search for an explanation of life's diversity, without considering design, is not only arrogant and foolish, but contrary to the tenets of the scientific approach.

Unfortunately, just as the ignorance and mysticism of the dark ages inhibited a proper understanding of the universe, so today, the inappropriate and scientifically disheveled and tattered dogma of biological evolution has inhibited a proper perspective in viewing the cosmos in which we experience our everyday existence.

For example, microbiologists never report a change in species, only a resistant strain of a known bacteria which has accommodated to certain antibiotics.

On close inspection, the evidence in support of macroevolution is speculative. No one has ever produced a functional new species from a pre-existing one, nor created life, or developed a family tree of evolution that is acceptable to most of us; nor have they factually explained the abrupt origin of thousands of diverse species, without precursors, during the Cambrian explosion (Cambrian explosion or Cambrian radiation has been described as a rapid appearance of most major groups of complex animals that happened about 530 million years ago). Those who taught this doctrine have not explained intuition, the sucking reflux, or a marsupial infant's instinct to climb into its mother's marsupial pouch to undergo further development, and so forth.

Another failure of Darwinism is the inability of the hypothesis to explain the method by which an "evolutionary" process, operating at the genetic level, might manipulate the molecules of a population's genome in such a way as to overcome the random chance base-pair point-mutations, insertions, deletions, and frame-shifts that are more than capable of overriding any selective process propagated by a preexistent reproductive or fitness advantage in an organism's phenotype.

COULD THE STEP-BY-STEP gradualism of Darwinism really produce integration of the motor and pre-motor cortices, the basal ganglia and the cerebellum, all on a stepwise purposeless stochastic process (stochastic is synonymous with random)?

In *The Origin of Species*, Darwin stated that if it could ever be demonstrated that an organ system could not be produced in a stepwise manner, his hypothesis would be disproved.

Darwin's theory of descent with modification (on the basis of natural selection) has been challenged by evolutionists themselves. They are divided on the question of whether macroevolution might occur by means of gradualism (neo-Darwinism), or by punctuated equilibrium (saltation theory of major mutations).

Evolutionist Richard Dawkins has stated: "It is grindingly, creakingly, crushingly obvious that if Darwinism was really a theory of chance, it couldn't work." Richard Goldschmidt, at the University of California, Berkeley, in the last century, saw "bridgeless gaps" in Darwin's descent with many small steps. In his 1940 paper, he postulated the hypothesis of large sudden jumps in the form of "hopeful monsters" (mutants with small changes in "control genes") that might produce large changes during embryogenesis. Stephen Gould defended Prof. Goldschmidt's concept when he stated: "Macroevolution is not microevolution extrapolated." In addition, the Hardy-Weinberg law forbids evolution in normally occurring populations. Godfrey Hardy, a mathematician, and Wilhelm Weinberg, a fellow physician, independently determined that the frequency of alleles, meaning any of a group of possible mutational forms of a gene, that is, a special form of genetic drift continue to produce the same variants throughout time. This remains the same in a population, generation after generation; even if one of the alleles is dominant.

When anyone tries to circumvent the Hardy-Weinberg law by using abnormally small populations with random drift, or bottleneck, or founder effect, that person runs up against Muller's ratchet (the increasing accumulation of deleterious mutations in small populations of asexually reproducing organisms, which can be rectified only by genetic death).

Evolutionists are unable to provide empirical evidence of inorganic materials being converted into life forms. The best they can do is to refer to the Urey-Miller experiment, which only proves that man can produce tar and a few amino acids — but not proteins and not life, by continually removing tar from the experiment, while running an electric spark through a reducing atmosphere in a glass container.

All of this speculation leads to poor logic.

Chapter Three
Evolution and Darwin

In trying to explain evolution, evolutionists can only postulate that a series of insertions, or deletions, point-mutations, or frame-shifts in the genetic code, might produce evolution of a species. However, such "mutations" are more likely to produce stop signals, or changes in the syntax of the genetic code, and thereby produce debilitation or death of an organism, not an improvement. That is the reason that almost all "mutations" are deleterious. No one ever purposely places pregnant females in a field of ionizing radiation, or gives them mutagens in the hope of producing an improved fetus.

THE MACROEVOLUTION CONCEPT of plant and animal evolution is a misleading illusion. Macroevolution is a phantom that materialized within the fog of ignorance that pervaded the 19th century. It has remained suspended in a mist of misconceptions even to this day. The preponderance of available scientific evidence demands that the diversity of life forms encountered on earth be considered a product of design – not chance.

Contemporary 21st century science reveals that Charles Darwin was wrong in every aspect of his basic understanding of, and his explanation for, the diversity of life encountered on this planet—as we shall explain in the chapters that follow. If we study the background and understanding of science that Darwin possessed, it is lacking in fundamentals.

Darwin was a disinterested medical student. He dropped out of medical school early in his career. In his autobiography, he stated: "Dr. Monroe made his lectures on human anatomy as dull as he was himself, and the subject disgusted me... I also attended regularly the clinical wards at the Hospital. Some of the cases distressed me a good deal. I cannot understand why this part of my medical course did not interest me in a greater degree; ... I also attended on two occasions the operating theater in the hospital at Edinburgh, and saw two very bad operations. I never attended again."

Thus, it becomes clear why Darwin chose to be a naturalist, a far less demanding pursuit, especially in view of the vagaries of 19th century biology. Being a naturalist required far less responsibility or precision of thought than did medicine or chemistry or physics or mathematics.

In this way, Darwin was able to flirt with the demands of science, as opposed to the rigors of developing scientific understanding. Darwin offers an explanation for his inability to pursue the demands of medicine in his autobiography: "I became convinced from very small circumstances that my father would leave me property enough to subsist with some comfort – my belief was sufficient to check any strenuous effort to learn medicine." This passage is particularly revealing.

There is another passage from Darwin's autobiography that is equally revealing about his intellect, and his ability to think logically: "During the three years which I spent at Cambridge, my time was wasted, as far as the academic studies were concerned, as completely as at Edinburgh and at school. I attempted mathematics, and even went during the summer of 1828 with a private tutor (a very dull man). I got on very slowly. The work was repugnant to me, chiefly from my not being able to see any meaning in the early steps in algebra. I did not proceed far enough at least to understand something of the great leading principles of mathematics; for men thus endowed seem to have an extra sense. I do not believe that I should ever have succeeded beyond a very low grade." He had difficulty in mastering algebra? Surprising!

Apart from his difficulty with logic, as required by mathematics, Darwin lacked an appreciation for the complexities of general anatomy, neuroanatomy, physiology, biochemistry, biophysics, etc. He was able to believe in chance evolution only because he knew so little about the interrelationships of the biologic functions that he was asking random chance and natural selection to integrate into the convoluted systems we refer to as life. Similarly, it is the same lack of a broad spectrum of scientific knowledge that continues to foster a belief in macroevolution.

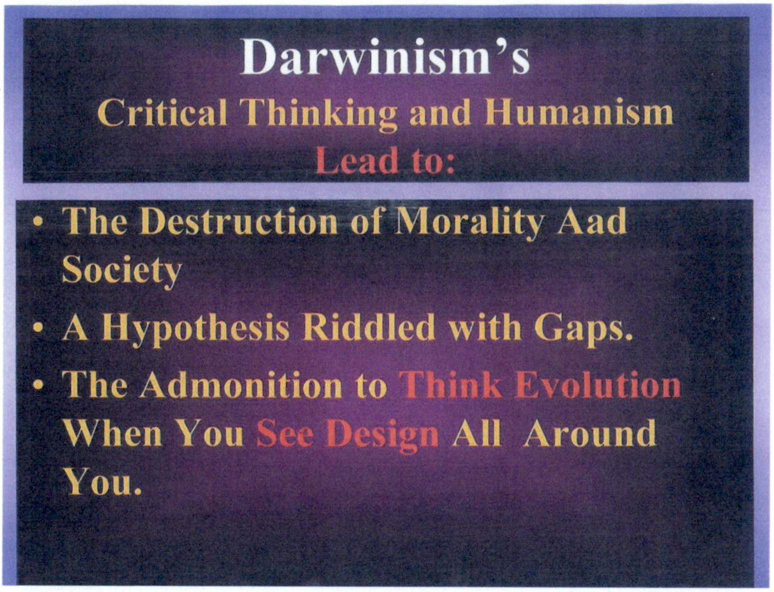

THE MATHEMATICS OF a chance origin of complex anatomic structures or physiologic processes forbids any chance/stochastic process as an explanation for life's origin and diversity. For example, the complicated neuromotor physiology required for smooth and fluid movements of running, jumping, swallowing, talking, chewing, or fixing random gaze, can be explained only on the basis of design.

A chance, layer by layer, build up of the neuromotor system is precluded. The amazing irreducible complexity of metabolism, fluid and electrolyte balance, pulmonary function, human fertilization, pregnancy, parturition lactation, etc., intuitively destroy the concept of chance origin.

In his unschooled thinking, Darwin knew nothing of genetics, or the work of Mendel. He erroneously attributed the changeable characteristics of succeeding generations of plants and animal variation to a dilution or melding of parental characteristics — the so-called paint-pot concept, i.e., the concept of changing white paint into various hues of the rainbow by blending a little bit of pigment from one source with that from another source (changing a phenotype by mixing varying concentrations of parental characteristics). Darwin knew nothing about alleles or dominant or recessive traits, or De Vries' postulate of phenotype quantum jumps on the basis of changes in parental "pangenes." In fact, Darwin knew very little about biology, by today's standards.

As a pigeon breeder, Darwin likened the production of life's diversity to an animal breeder's selecting certain genetic qualities for reproduction. He considered "Nature" to be the master breeder using a process Darwin termed "natural selection", to direct life into ever-higher domains of function and diversity (never mind the absence of available chromatin).

Darwin's hypothesis of "natural selection" was a spurious postulate from the beginning. Hugo DeVries, the Dutch botanist who rediscovered Mendel's laws of inheritance and coined the term mutation, precociously summarized the problem with Darwin's concept poignantly and completely when he pointed out that Darwin's postulate of "natural selection" might explain "survival of the fittest," but it couldn't explain the "arrival of the fittest."

In fact, the "arrival of the fittest" still cannot be explained in Darwinian terms, as you will see as we move along. The limitations of chance mutations and chance recombination, and the paucity of chromatin required for an evolutionary ladder, all contradict the concept of evolution.

MOREOVER, IT IS now patently obvious that "natural selection" is no longer a tenable scientific concept (we will discuss the science later), that has demanded evolutionists replace natural selection with the default concept of "neutral mutation with random drift" (Motoo Kimura).

The concept of natural selection just doesn't hold up against the fact that "mutations" accumulate in a species' genome, and are not selected out. (Between 14 and 30 percent of any given genome's alleles are polymorphic). To modify the concept of natural selection to mean reproductive success as opposed to a mechanism capable of producing selection does not alter the facts.

Today, the concept of natural selection has no basis in hard fact. DeVries was right. Darwin was wrong and so are current day evolutionists who still cling to a concept which is absolutely impossible to apply at the level of molecular biology. Why impossible? Because natural selection, as a hypothesis, can act only upon an existing phenotype; not a chance-directed intracellular molecular process.

Natural selection would have no way of exerting any effect on which mutations might, or might not, occur in a genome (all of the genetic information, the entire genetic complement, all of the hereditary material possessed by an organism), or where, or when, even if natural selection were a reality. A successful phenotype could be selected in one generation only to have its advantage wiped out by a chance mutation in the next.

The evidence cited above should be enough to dissuade evolutionists from their blind faith in the siren they call evolution. However, there is even stronger, irrefutable, noncontributory evidence from the discipline of statistical chemistry.

Large polymeric molecules (a large molecule made up of chains or rings of linked monomer units) produce a family of polymeric molecules, though not one specific type, as they polymerize. DNA is a large polymeric molecule; the larger the molecule, the larger the error-rate. Therefore, a specific type of DNA molecule cannot be maintained without the aid of special enzymes (DNA-polymerases) that maintain fidelity. Left to the laws of statistical chemistry and molecular biology, natural selection could never select out any one specific DNA phenotype (the observable physical or biochemical characteristics of an organism, as determined by both genetic makeup and environmental influences).

THIS PHENOMENON OF PHENOTYPES is known as Eigen's Paradox. Evolutionists Stephen Stearns and Rolf Hoekstra summarized the Paradox in their textbook of evolution as follows: "Irrespective of the precise conditions of prebiotic chemical evolution, a collection of organic compounds does not imply life. We do not yet know how RNA could be formed in the prebiotic environment."

If errors in replication occur with a constant probability per subunit, then long molecules are replicated with more mistakes than short molecules. It can be shown that a critical size of a replicator molecule exists (the so-called error threshold) above which larger molecules cannot be maintained because they are replicated with too many mistakes. Eigen's Paradox then follows:

"Non-enzymatic replication has low accuracy, so that only small molecules can be maintained for accurate replication enzymes needed, but the primitive genomes were too small to code for them. This is the catch-22 of prebiotic evolution: large genomes are only possible with replication enzymes, but replication enzymes require large genomes."

Disregarding the lack of an explanation for the mechanism by which a DNA/RNA (RNA is a polymeric constituent of all living cells and many viruses; it consists of a long, usually single-stranded chain of alternating phosphates) genotype/phenotype can be maintained, evolutionists have turned their attention to the mechanism by which new species might arise, i.e. mutation, a concept of which Darwin, himself, was unaware.

Today, the term mutation continues to convey an incorrect concept. Many "mutations" are nothing more than changes in nucleotide sequences that occur during DNA replication, or recombination. If a mutation strays too far from normal, and produces a significant insertion, deletion, or frame shift, the result is a change in the syntax of the gene coding, with resultant hypoactive or inactive enzymes.

Mutations, in the true sense of the word, can be produced with mutagenic agents, but not in "wild" type species in nature. No one, anywhere, has ever converted one wild type species into another, or into a new, functional species.

The genetic changes that evolutionists refer to as mutations are simply biomolecular variations that you and I, and any rationally thinking person trained in science, would call normally occurring statistical variations, some permitted and others not permitted, and anticipated within the limits of cellular biology and genetic variability – designed to allow a given species to accommodate to a changing environment, and to do this without being converted into a new species. A mutant that strays very far from a typical design rarely survives.

Worse yet, mutations are subject to Mueller ratcheting, which is defined as a problem that occurs in small populations of asexually reproducing organisms in which mutation rates, per force, stay the same or increase. The mutations can be eliminated only by genetic death. Stearns and Hoekstra point out that if the mutation rate per zygote reaches 1.0 or higher, the genome of the population in question is in serious trouble. The current estimated mutation rate in human beings is 1.6 per zygote.

This would appear to be a serious problem for evolutionists who try to circumvent the Hardy-Weinberg law using small populations and genetic drift.

In light of the above discussions, we see that not even an amalgamation of the concepts of Darwinism, mutation, and genetics – the so-called modern synthesis/neo-Darwinism – have been able to salvage Darwin's untenable hypothesis, except in the minds of those who want to believe and who choose to disregard the complexities of neurophysiology, the chicken and egg relationship between DNA, RNA, protein and Eigen's Paradox, and the fact that 99.9 percent of all mutations are either deleterious or fatal. (The internet encyclopedia, Wikipedia, uses 70 percent as the portion of mutations that are harmful to a species. I have used 99.9% because in medicine, we take no chances.)

TO BELIEVE IN EVOLUTION is to speculate and to hope against hope, to disregard fact and logic, and to adhere to philosophy rather than science. There is but one logical answer and only one logical answer capable of explaining the reliable replication of large molecules and the maintenance of various species and their differences. That answer is the designed enzymatic shaped-space of enzymatic proteins, i.e. preexistent enzymes.

If evolution were true, there would be a multitude of trees of life, not one discernible single tree. Homology (the quality of being similar or corresponding in position or value or structure or function) is evidence of design, not chance evolution. We will, also, compare the overwhelming evidence that points to design as the causative factor producing life's diversity, as compared to the dead-end retrograde extrapolation of chance "evolution." We will allow that chance evolution can produce limited microevolution (intraspecies variation), but not macroevolution (a change in species).

WHY DOES EVOLUTION continue to be taught? It is instructed in school as though it were fact, in spite of the overwhelming evidence to the contrary. It appears that the primary reason is that those who believe in Darwinism are those who are limited in their understanding of the broad expanse of science.

They fail to understand the demands of neuroanatomy, the demands of biped ambulation, bioengineering, biomanufacturing, the biochemical and biophysical facts, and a myriad of other considerations. Those who expound evolution feel secure within the limitation of their abstractions, and the boundaries of the paradigms in which they live and think. They are unaware of the scientific facts that contradict the reality of the hypothesis in which they put their faith, and are unwilling to follow reality and its trail of facts — wherever reality's path might lead. It is time to concentrate on design in the Solar System, and why biologic design on Earth requires so many variations.

Chapter Four
Ascendancy of Intelligent Design

A hydrocarbon–based body, alone, is not the essence of our reality. Quantum reality imparts a dual existence to our lives; and life's various manifestations are better thought of on a functional basis, as a myriad of diverse energy transducers — all designed to maintain ecostasis (against bacteria, fungi, molds, etc.), with some specifically designed to convert electromagnetic energy (sunlight) into stored potential energy (plants), and some specifically designed to convert stored potential energy into kinetic energy (animals). Others are designed to recycle the original natural resources expended in the production and function of each of these instruments of ecostasis.

All of these are working together under the influence of a governing negative feedback system which maintains the biosphere that supports life as we know it. Life's transducers can undergo modification to accommodate to changing environments, within design limits, but one cannot be made to change into another. That is a fact.

This concept of a balanced and integrated negative feedback is diametrically opposed to the Darwinian concept that life-forms are independent, self-contained, accidentally derived competing entities, born into a senseless and arbitrary amalgam of natural laws and elements, which spend their lives utilizing fuel and resources in an endless pursuit of self-preservation.

THE BETTER EVIDENCE-BASED explanation for life is that plant and animal bodies are elegant homeostatic, counterbalancing, independent subsystems, interconnected, and symbiotic, yet self-serving and self-maintaining. They are like interacting factories, or automatons, that take in substrate and contemporaneously convert raw material into motion, activity, nutrients, shelter, cell division, computers, tires, wheels, conveyor belts, boilers, generators, engines, electrical systems, transducers, free energy, etc.

All of these are constantly changing with chemical interchangeabilities that are in a constant dynamic, negative feedback equilibrium with each other, and with the environments in which they find themselves.

At the same time, these plant and animal bodies maintain their general outward appearance, as though their "phenotypic" characteristics were static – while in reality, inwardly, functioning electrochemically ("genetically") in a dynamically interactive symbiosis with the environments in which they find themselves, as modified by built-in reflexes, instincts, and intellect, etc.

Study the amazing human body. Every day, our bodies break-down and rebuild glycogen and glucose molecules, fatty acids, amino acids, proteins, and cells of all types. They use their carbon skeletons to convert one substance into another, and back again. All of this by means of anaerobic glycolysis, gluconeogenesis, oxidation, beta-oxidation, deamination, phosphorylation, and other metabolic processes.

This constant transition results in energy production, energy storage, cell maintenance, and mental and physical activity — all, primarily, using the high energy phosphate bonds of adenosine triphosphate (ATP) as the preferred final common pathway of energy transfer (except for the metabolism of the brain). Could this symbiotic interchange have come about by chance? Maybe! Read on before deciding.

It is an amazing event to need a burst of energy and to have the adenosine triphosphate (a nucleotide derived from adenosine that occurs in muscle tissue) be there as the major source of energy for cellular reactions ready to provide energy at the rate of 50 kcal per minute (500 grams per minute — for a period of 10 to 15 seconds).

The ATP expended, can then be regenerated, or, additional energy can be immediately provided by anaerobic glycolysis (the breakdown of glycogen to glucose 6-phosphate) at the rate of 30 kcal per minute. The anaerobic glycolysis so required is then limited by the production of lactic acid. But this impediment is controlled by the lactic acid being converted to glycogen in the liver, or being oxidized to carbon dioxide and water in the Krebs citric acid cycle – which allows a constant aerobic energy production rate of 12 kcal per minute – over extended periods of time.

One or the other of these methods of energy production and transfer might occur by accident; but what about the occurrence of all of them being woven together, in dynamic equilibrium?

BOTH ORGANIC AND inorganic equilibrium constants favor the production of certain molecules, under specific conditions. But, what combination of the laws of nature is capable of "deterministically" setting up coexistent, but independent, and integrated, self-maintaining, interactive chemical systems, systems with individual equilibrium constants, and this within a single milieu, when the unbridled laws of chemistry demand that there be but one equilibrium constant for any given overall system with one final, dead-end, concentration of products and reactants, determined by the parameters delineating the system. The interactive combination of such energy producing systems is prohibitively unlikely, except on the basis of design.

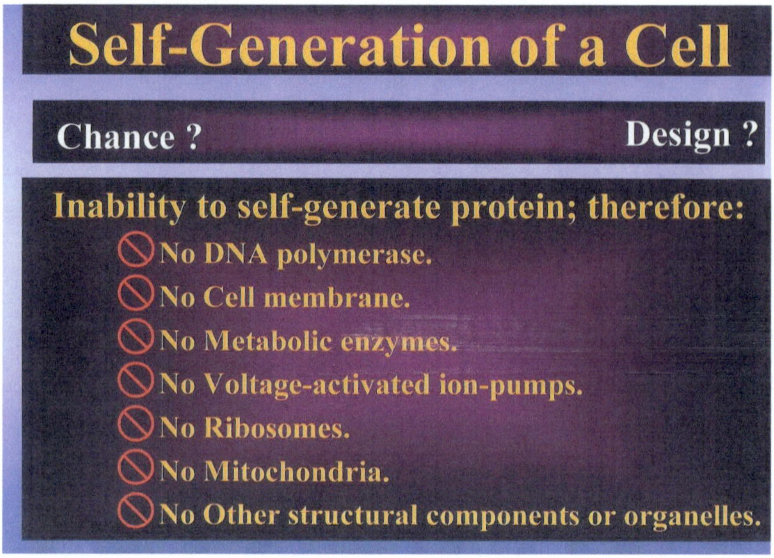

With the above in mind, note that on a daily basis, the standard 70 kilogram human being turns over the carbon chains that constitute his or her body in sufficient amount to produce, and consume, approximately, his or her own body weight of adenosine tri-phosphate (63 kilograms that represent approximately 140 pounds of ATP, with a minimum production of 45 kg per day in resting individuals).

All of this in order to power the energy needs of cell maintenance, cell reproduction, energy absorption, energy utilization, the conversion, and re-conversion of glucose, fatty acids, and non-essential amino acids into one another. In addition, the production of initial-chemical-activation energies, and the energy needed for human mechanical and electrical activity. In the presence of physical activity, ATP utilization can be as great as 500g per minute.

Moreover, there is an allosteric molecular switch, phosphofructokinase, which inhibits the Krebs cycle when the concentration of ATP is high, and activates the system when ATP is low, or when its precursor adenosine diphosphate (ADP) is high. An allosteric molecule is one which can exist in two or more chemically active configurations, depending upon the concentration of other molecules. What a system! What a switch! What a design!

The elements of design are obvious, and easy to see. But who is ready to stand up and empirically demonstrate how all this could have come about on the basis of meandering, accidental happenstance?

It is not for those of us, who believe in design, to explain how an accidental occurrence might have come about. But, rather, it is for those who want to believe in non-goal-directed evolution to explain such an occurrence – step-by-step and, convincingly, prove each step. In the meantime, until a factual demonstration of accidental occurrence is brought forth, the existing evidence will remain evidence in favor of design—not evolution. The burden of proof lies with evolutionists.

WHICH THEORY BETTER explains the spectrum of life-forms we observe on this planet, intelligent design or chance Darwinian macroevolution? Equally intelligent, but not necessarily equally knowledgeable, students of nature hold sharply opposed answers to that question. As we shall demonstrate, the available evidence overwhelmingly favors design, intelligent design.

Supporters of a chance origin of life and subsequent evolution mistakenly believe that the finite amount of genetic information available in simple organisms can be rearranged to produce organisms of ever increasing complexity. The laws of statistical chemistry prohibit such increases in complexity, even in an open system.

The reason is the absence of an independent method for producing chromatin, i.e. DNA. A cell can only duplicate the chromatin naturally inherent within its cell membrane, molecule by molecule, sequence by sequence.

A cell cannot produce new, virgin chromatin, denovo, then mold that chromatin into genes or chromosomes. In addition, denovo chromatin production is subject to a chicken and egg relationship between DNA, RNA, protein, and Eigen's Paradox. Life cannot exist without protein to support enzymatic reactions, structural tissues, internal organs, etc.

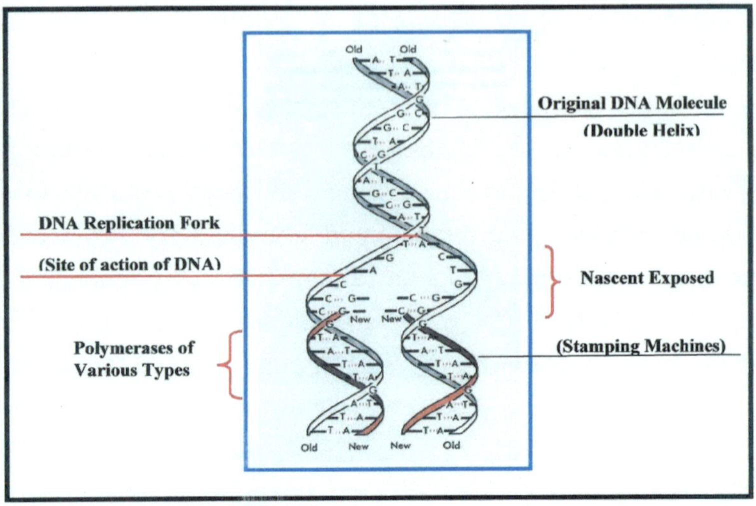

Elements of an Information Induction System:

The diagram represents the unzipping of a DNA molecule by DNA polymerase by disruption of the hydrogen bonding of the adjacent base pairs; thus forming nascent exposed bases within the shaped space of DNA polymerase. The substrate exposed to nascent bases then reform hydrogen ion bonding with their complimentary base pairs to reconstitute exact duplicates of the original molecule. It is obvious that this method or reproduction reproduces essentially exact duplicates with geometric reproduction speed. The method of production for RNA is essentially the same except the process occurs in transcription bubbles of DNA, and only one of the exposed helices is used to form the transcription of RNA.

Protein production requires the presence of DNA and RNA, both of which require the proteins DNA-polymerase and RNA-polymerase for their production – accordingly the chicken and egg relationship.

There is also a chicken and egg relationship (which came first) that contradicts a spontaneous origin of life and evolution apart from the denovo formation of chromatin. The contradiction is Eigen's Paradox, which states that: "large genomes are only possible in the presence of replication enzymes but replication enzymes (proteins) require large genomes."

Replication errors increase in a polymeric system in direct proportion to the increasing length of the polymers in question — in the absence of specific enzymes. DNA is a very long polymer. This is why the probability of reproducing DNA with fidelity, in the absence of the catalytic space provided by DNA-polymerase, becomes prohibitive. Prohibiting the chance origin of reproducible life forms in the first place.

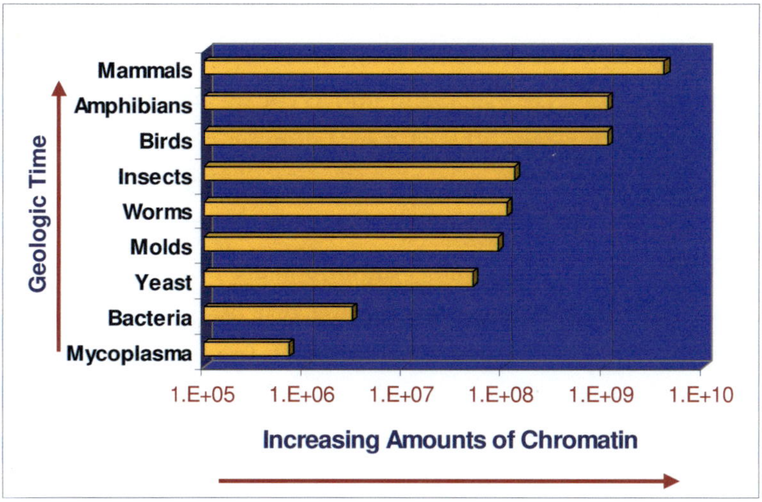

Relative Amounts of Chromatin/DNA per Phylum. The minimum genome's size (and amount of nuclear chromatin) found in each phylum increases from prokaryotes to mammals. (Redrawn from B. Lewin: Genes II: Oxford U. Press, N Y; 2002)

Moreover, the rearrangement of existing finite information within a population's genome doesn't produce any new information or increase in genetic complexity. All that one receives from such random mixing is a recombination of old information, with the same limits on complexity.

Additional chromatin is required to produce increases in information and complexity. But where would additional chromatin come from? Chromatin is just another name for the polymeric DNA molecules, which constitute chromosomes. Thus chromatin remains under the constraints of statistical chemistry, as already discussed before.

Could "mutations" provide the additional chromatin? No. First of all, mutations, as described in molecular biology, are nothing more than simple base-substitutions, or genetic insertions or deletions, or as recognized in cellular biology, mutations are nothing more than chromosomal recombination of existing chromatin. Secondly, the confines of the catalytic shaped-space of DNA replication forks and RNA transcription bubbles do not allow the type of saltation mutation jumps postulated by DeVries, and hoped for by evolutionists.

In medicine, we think of a mutation as producing "mutants" – catastrophic departures from the norm — caused by ionizing radiation, or by mutagens such as rubella (German measles), etc.

With some, evolutionists mutations are mistakenly regarded as minor, random variations in the normal process of cellular reproduction, variations capable of producing small, random, stepwise, benign changes in a population's genome, mythically leading to an overall improvement in the species, and eventually to new "higher" species, all by means of chance "mutation" and "natural selection." The concept is absolutely contrary to logic, probability, and scientific fact.

The problem with the hypothesis is fourfold. First, there's no method for the production of increased amounts of chromatin. Secondly, the chicken and egg relationship between DNA, RNA and protein, along with Eigen's Paradox, prevents the formation of new species. Thirdly, significant mutations are bad for an organism, and normally cause death, when significant. And lastly, any randomly occurring variations in DNA molecules, within the limits of allowable physiologic change at the molecular level, can never be made subject to a selection process.

ORIGINALLY, DR. GODFREY H. HARDY, a mathematician, and Wilheim R. Weinberg, a physician, independently recognized that the laws of statistics, and Mendelian inheritance, demanded that the frequency of alleles remain essentially the same in each succeeding generation of a population.

In its essence, the Hardy-Weinberg law states that population curves, subject to the laws of inheritance, cannot be shifted by selection, Erwin Schrodinger, the father of molecular biology, was very aware of this fact when he stated, and I paraphrase: "Darwin was mistaken regarding the small continuous, accidental variations, that are bound to occur even in the most homogeneous population as the material on which evolution works." This leads to the concept of descent regarding Darwinism.

Chapter Five
Irreducible Complexity

The concept of *irreducible complexity,* as it has been introduced, means that a single system which is composed of several interacting parts that contribute to the basic function of the system, and where the removal of any one of the parts causes the system to effectively cease functioning.

Michael J. Behe is an American biochemist and intelligent design advocate. He currently serves as professor of biochemistry at Lehigh University in Pennsylvania and as a senior fellow of the *Discovery Institute's Center for Science and Culture.* Behe is best known for his argument for irreducible complexity. He asserts that some biochemical structures are too complex to be adequately explained by known evolutionary mechanisms, and are therefore more probably the result of intelligent design.

Supporters of intelligent design use this term to refer to biological systems and organs that they believe could not have come about by any series of small changes. They argue that anything less than the complete form of such a system or organ would not work at all, or would in fact be a *detriment* to the organism, and would therefore never survive the process of natural selection.

Although they accept that some complex systems and organs *can* be explained by evolution, they claim that organs and biological features which are *irreducibly complex* cannot be explained by current models, and that an intelligent designer must have created life or guided its evolution. Accordingly, the debate on irreducible complexity concerns two questions: whether irreducible complexity can be found in nature, and what significance it would have if it did exist in nature.

IF WE PONDER THE SIGNIFICANCE of design in our existence, the thought begs yet another question: if our existence is by design, then who, or what, produced that design and for what purpose? Although the argument pursued within this treatise is limited to the controversy between design and accidental evolution as the explanation for life's diversity, nonetheless the more information, and the more discussion that can be directed at the subject, whether from a scientific stand point, or a philosophic stand point, or a moralistic stand point, the better will be society's collective understanding of the essence of what we call life.

If this concept is adhered to, then the better will be the chance that those possessing such understanding will conduct themselves as intelligently designed humans, rather than wanton animals. Thus, as stated previously, the author is compelled to share the scientific and philosophic information gathered, through a lifetime of medical endeavor, with others — for their evaluation, and perhaps use as a steppingstone in developing their own perspectives on life — and, if for nothing else, as general information.

Ignoring fact is foolish – especially scientific fact. To do so is to indulge in self-delusion. There is no reason to deceive ourselves as to who we are, or why we exist at all. As rational human beings, it is important that we devote time to addressing the most important judgment we will ever make. Do we exist by chance, or by design? There are many examples of empirical evidence that refute macroevolution from the standpoint of probability.

In the real world of everyday life, when we see a complex machine, we do not intuitively believe that it sprang full-blown from the forehead of Minerva. Rather, we know from experience that complex machines are developed over time with successive specifically-designed refinements. Thus, the computer on which this treatise was typed started out as a simple calculator which was modified and refined to play Mary Had A Little Lamb, and subsequently into the amazing machine which can now navigate the Internet, draw the illustrations found herein, change the line-spacing, justify the margins, change the font to any color, and carryout multiple tasks at once, etc.

Darwin very well knew that he was contradicting the human experience of everyday life when he postulated that the complex organic mechanisms that we take for granted, and see everywhere, came about by chance – if for no other reason than he was unable to convince his own wife that the human eye was the product of successive small changes that occurred over time on the basis of natural selection. She was right to be skeptical.

In an effort to circumvent logic and everyday experience, Darwin in his day, and evolutionists such as Richard Dawkins, today, resorted to an unallowable and factually contradicted scheme. They would climb the un-climbable Mount Improbable using the gently sloping paths along the gently sloping contours of the backside of Mt. Improbable, rather than trying to scale its steep face from the zero-baseline to the pinnacle of its improbability in one quick jump. In the minds of persons like Darwin and Dawkins, such a scheme seemed plausible – but only because they failed to consider all the requirements, and all the implications.

"In Darwin's thinking, evolution could not build a complex organ in one step or a few steps; radical innovations such as the eye would require generations of organisms to slowly accumulate beneficial changes in a gradual process. He realized that if in one generation an organ as complex as the eye suddenly appeared, it would be tantamount to a miracle.

Unfortunately, gradual development of the human eye appeared to be impossible, since its many sophisticated features seemed to be interdependent. Somehow, for evolution to be believable, Darwin had to convince the public that complex organs could be formed in a step-by-step process. He succeeded brilliantly. Cleverly, Darwin didn't try to discover a real pathway that evolution might have used to make the eye. Rather, he pointed to modern animals with different kinds of eyes (ranging from the simple to the complex), and suggested that the evolution of the human eye might have involved similar organs as intermediates."

Professor Behe hit the nail precisely on the head when he used the term "cleverly", and pointed out that Darwin did not try to "discover a real pathway" to explain evolution. Darwin simply offered up a conjecture and left the determinist philosophy of his age do the rest.

Professor Dawkins is not a biologist. Rather, he is an ethologist - one who deals with speculation, non-verifiable concepts, and philosophies regarding animal instinct. His book clearly reflects the imprecision of this thought processes. In reality, professor Dawkins appears to subscribe more to the thinking of a humanist and a philosopher, rather than a scientist.

DARWIN WAS NOT TRAINED in the exacting requirements of science either. Thus, both natural selection and cumulative selection are seen to be what they are, the products of fertile imaginations and a desire to explain nature in the simplistic, fairy-tale rubric of "kiss a frog, wait 10,000 years, and the frog will become a prince," while sidestepping the demands and rigors of scientific investigation and deduction and the constraints of the bounds of probability.

We all owe our thanks to Professor Behe, not only for exposing natural selection and cumulative selection for their non-scientific origins, but also for formalizing the concept of irreducible complexity. He defined irreducible complexity as follows: "By irreducible complexity I mean a single system which is composed of several interacting parts that contribute to the basic function, and where the removal of any one of the parts causes the system to effectively cease functioning. An irreducibly complex system cannot be produced directly by slight successive modifications of a precursor system, since any precursor to an irreducibly complex system is by definition nonfunctional." (Pennock or Dembski Debating Design.)

Chapter Six
Mathematics and the Falsity of Evolution

Darwin proposed the criterion that would discredit evolution. He put it this way: “If it could be demonstrated that any complex organ existed which could not possibly have been formed by numerous, successive, slight modifications, my theory would absolutely breakdown.”

Irreducible complexity in biologic systems demonstrates just that. As discussed previously, professor of biochemistry, Michael Behe, was the first to formalize the concept of irreducible complexity as follows: “An irreducibly complex system cannot be produced directly by slight, successive modifications of a precursor system, since any precursor to an irreducibly complex system is by definition nonfunctional.”

Let’s see if we can employ the irrefutable power of mathematics to prove that some biologic systems are irreducibly complex and could not have been produced in a stepwise manner, and thus falsify the concept of macroevolution, using Darwin’s own criterion.

The Chicken and Egg Dilemma

The chicken or the egg causality dilemma is commonly stated as "which came first, the chicken or the egg?"

Chickens hatch from eggs, but eggs are laid by chickens, making it difficult to say which originally gave rise to the other. To ancient philosophers, the question about the first chicken or egg also evoked the questions of how life and the universe in general began.

Cultural references to the chicken and egg intend to point out the futility of identifying the first case of a circular cause and consequence. It could be considered that in this approach lies the most fundamental nature of the question. A literal answer is somewhat obvious, as opposed to the logical fallacy of the metaphorical view, which sets a metaphysical ground on the dilemma. So, to understand its metaphorical meaning better, it could be reformulated as follows: "Which came first, X that can't come without Y, or Y that can't come without X?"

WHEN TRYING TO EXPLAIN THE ORIGIN OF LIFE, or the replication of proteins, evolutionists run up against the chicken and egg (which came first) relationship between DNA, RNA and protein. You can't make DNA without protein (DNA-polymerases); and you can't make protein without DNA and RNA (and protein RNA-polymerases). They also encounter Eigen's Paradox, which, in essence, states that it takes the catalytic space of replication enzymes (DNA-polymerases) to produce large genomes; but it also takes large genomes to produce the catalytic space of the replication enzymes.

There is also the problem that the overwhelming number of naturally occurring mutations is detrimental; the constraints of the fourth law of probability; and the allowable extremes of probability bounds.

Let's consider the chicken and egg relationship between DNA, RNA, and protein. Protein cannot be produced without RNA. RNA cannot be produced without DNA. DNA cannot be produced without proteins, i.e. DNA polymerases. Thus, we have an irreducible chicken and egg relationship. This is an example of irreducible complexity that appeals to the intellect. However, the physiologic processes involved in the relationship defy mathematical quantification. Thus, this relationship does not lend itself to mathematical analysis.

Just as with the chicken and egg relationship between DNA, RNA and protein, there are multiple intuitive examples of irreducible complexity in the human body. Unfortunately, few lend themselves to easy mathematical analysis.

In some cases, the mechanisms involved are not well enough known. In other instances, the modes of action are simply too complex for proper interpretation. However, there are two systems that do lend themselves to mathematical evaluation using the known laws of physics and mathematics. One is the effect of surface tension on the alveoli of the lung, and the other is the application of the classic laws of probability to the chance integration of the motor cortex and basal ganglia of the brain.

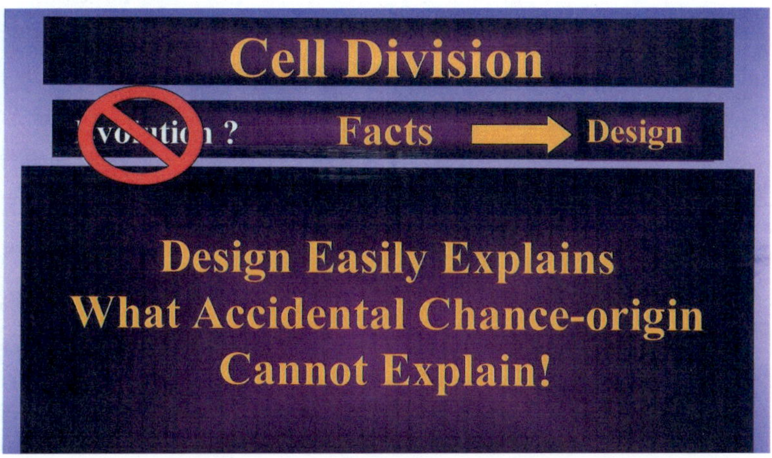

In order to function, the mammalian lung requires not only the construction of hundreds of thousands of small alveoli, and myriads of blood vessels, but the improbable production of a special group of phospholipid surfactant molecules: primarily dipalmitoyl phosphatidylcholine, along with five other phospholipids, and four apoproteins.

These special molecules can both adjust surface tension and resist compression. It is the presence of these molecules that allow us the luxury of relatively small lungs without the severe energy consumption problems that would normally be encountered in the presence of small diameter alveolae, air-liquid boundary layers, surface tension, and the need for great trans-pulmonary force in moving air in and out of the thorax.

THE FOLLOWING SCENARIO then becomes a standard. A gene undergoes a mutation, one that may facilitate the reproduction of those individuals carrying it, overtime, and with respect to a specific environment, mutants come to be statistically favored, replacing individuals lacking the requisite mutation. Evolution could not be an accumulation of such typographical errors. Population geneticists can study the speed with which a favorable mutation propagates itself, under the circumstances.

They do this with a lot of skill, but they are academic exercises if only because none of the parameters that they use can be empirically determined. A typographical change in a computer program does not change it just a little. It wipes the program out, purely and simply.

Now, speaking of macromutations, let me observe that to generate a proper elephant, it will not suffice suddenly to endow it with a full-grown trunk. As the trunk is being organized, a different but complementary system, the cerebellum, must be modified in order to establish a place for the ensemble of wiring that the elephant will require to use his trunk.

These macromutations must be coordinated by a system of genes in embryogenesis. If one considers the history of evolution, we must postulate thousands of miracles; miracles, in fact, without end. No more than the gradualists, the saltationists are unable to provide an account of those miracles. No algorithm allows us to understand the complexity of living creatures.

Design parameters require a large surface area for proper aeration of the lungs and proper exchange of oxygen and carbon dioxide.

Contrary to the methods by which insects and birds satisfy their oxygen requirements, mammals use a system of expansion and contraction of the thorax, which moves air in and out of hundreds of thousands of alveoli of variable size.

Anyone who has ever blown up a balloon understands that the initial degree of inflation of a balloon determines the ease with which it can be filled with air. At first, in its collapsed state, it is difficult to force air into the balloon. However, as the balloon expands, it becomes easier and easier to force more and more air into the balloon until it finally pops. The reason is that the force per unit area, acting outward from within the balloon, remains the same; but, the surface area increases as the balloon expands. Therefore, the total expansion-force within the balloon is ever increasing, because the total force within the balloon is a function of the total surface area.

In mammals, expansion of the thorax creates a negative transthoracic pressure, which expands the lungs and their alveoli. Thus a gradient is produced between atmospheric pressure and the negative transthoracic pressure within the thorax. Under normal conditions, the lung expands and contracts easily for two reasons: (1) the alveoli are not allowed to collapse, and (2) the physical effect of surface tension is modified, as will be explained below.

Surface tension exists wherever a gas/liquid interface exists. Surface tension is what allows bubbles to form, and certain insects to walk on water. Because the alveoli are coated with moisture, the laws of physics and surface tension, at the air/liquid interface, demand that the effect of surface tension on alveoli, of different sizes, should cause air to flow out of small diameter alveoli, and into larger diameter alveoli.

Even if the effect of surface tension were lessened, by the use of a simple wetting agent (such as plasma), the general physical effect would still hold true. Because of the mathematical laws that govern surface tension and the diameter of spherical, elastic cells such as alveoli, the air would still flow from small alveoli into larger alveoli, causing the smaller alveoli to collapse, and the larger alveoli to over-expand – thus producing areas of atelectasis and emphysema in all lung tissue.

Can any intelligent person believe that random chance could have produced such an array in the right place at the right time in order to accommodate mammalian lungs, or the lungs of any other animal? What is even more unbelievable is that surfactant is also capable of resisting compression, so as to strongly resist complete collapse of an alveolus, and atelectasis.

If it were not for surfactant, there would be no normal alveoli, only atelectasis and emphysema.

THE CLASSIC LAWS OF PROBABILITY

- First Law: If the probability of an event is 1, then the event must occur.
- Second Law: If the probability of an event is 0, the event will not occur.
- Third Law: The probability of an event occurring, at all, lies somewhere between 0 and 1.
- Fourth Law: The sum of all the probabilities of all simple events which might occur in a sample space must equal 1, and cannot be greater than 1. Whenever something can have more than one outcome, which are equally likely, then the probability of any particular combination of outcomes on two or more independent trials is the product of their respective probabilities. (Note: with trial and error stochastic methods, the probability of success or failure, with each "mutation," is equal.)
- Sample Space: A sample space is a compilation of all possible outcomes that might occur, given a set of parameters. For example, given 12 plates of different colors, and a table on which to set the plates, there are 12 ways in which the dishes

can be arranged around the circumference of the table, starting at the 12 o'clock position.

- Odds Ratio: The odds-ratio is determined by dividing the probability that an event will happen by the probability that the given event will not happen.

When determining probability, infinite time, infinite energy, infinite methods of production, stochastic vectors, degrees of belief, subjective interpretation and inductive logic do not apply, and neither do the mathematics of information theory.

Part II

The Complexities of the Human Body

Chapter Seven
Male and Female Reproduction

Multiple different types of cells, proteins, tissues, and methods of function have to be integrated to produce the structures and chemicals necessary for life. If evolution were true, it would require multiple simultaneous chance mutations on multiple genes, in various locations along various chromosomes, in order to produce the integrated complexity required for transitioning function – not only with regard to gross anatomy, but also with regard to microscopic anatomy, physiology, and biochemistry, i.e. the biochemical production of special chemicals such as surfactants, etc. To the scientifically trained mind, this is all evidence of design, not chance evolution.

In the maintenance of life, as described in a veterinary textbook of clinical anatomy and physiology: "The whole, grand purpose of the male and female reproductive cycle is fulfilled when parturition, the birth process, successfully takes place. At birth, the fetus, which has led a parasitic existence of until now, is pushed out of the warm dark, moist environment of the uterus into the cold, cruel outside world as a free living independent animal. In a very short time, the body of the fetus has to undergo some pretty dramatic changes. Its lungs, which have been nonfunctional until now, must suddenly expand and start functioning to supply the newborn with the oxygen it needs and eliminate the carbon dioxide its cells are producing as a waste product."

"Contributing to this process are some dramatic changes in blood flow in and around the heart and lungs. The foramen ovale and ductus arteriosus which shunted most of the blood from the right side of the fetal heart around, rather than through, the lungs must both close fairly quickly. The lungs now need their full blood supply to properly oxygenate the blood and eliminate carbon dioxide.

Precisely what triggers parturition is not known. Many factors seem to be involved, including the size and weight of the uterus and fetus and changing hormone levels in the fetus and dam. The hormonal changes form a chain of events that leads to the onset of labor. Around the time of labor, the progesterone level in the bloodstream of the dam declines."

The foregoing summary of the high points of mammalian gestation and parturition should be sufficient to exemplify the tremendous complexity of the system. However, that is only the beginning, because a discussion of the maintenance of life, over time, is even more complex.

A discussion of the entire system includes discussing: the instinctive desire to mate and nest, the desire for sexual intercourse, insemination, fertilization, gestation, parturition, the sucking reflex and postpartum infant nutrition, transfer of immunity, neonatal development, and independent growth and life.

FOR THE SAKE OF ARGUMENT, let's take the accidental origin of life as a given — with the understanding that such is impossible because of the chicken and egg relationship between DNA, RNA, and protein, and the statistical considerations of statistical chemistry, which are the problem with having more than one equilibrium constant in the same location in the same primordial soup, at the same time, and the laws of thermodynamics.

Nonetheless, accepting the chance origin of life as a given, and also accepting the tenets of neo-Darwinism: that with genetic mutation, one thing leads to another, and natural selection selects out the best resulting candidates on the basis of reproductive success, we can further postulate that "evolution" might have produced the diversity of life that we now see on the earth (again with a caveat that natural selection is a concept which has been discarded in favor of the neutral mutation with random drift theory of Motoo Kimura).

Reasoning from the point of view of these parameters having been postulated as givens, we must yet include in our assumption that chance origin provided us not only with life, but with a self-sustaining system, capable of reproducing itself, and not only reproducing itself, but "evolving" into other forms of life.

WHY DO MEN DESIRE WOMEN? Why do they have such a strong desire for intercourse? Because without that strong desire and frequent intercourse, the chance of the female becoming pregnant is very poor, as we shall discuss later. What if men and women had lived independently, or together, simply enjoying one another's company only with a platonic relationship? What would have happened to the species? It would long ago have died out.

No, it is incumbent on the theory of evolution to build not only new phenotypes from the genomes of old populations, but also to build both male and female prototypes of each new sexually reproducing species, and build them at the same place, at the same time, and with the proper mating instincts. It would do no good to have the male of a new species produced in New York, by chance, and the female of a new species produced in California, by chance. And, it would do no good if they didn't want to copulate, even if their chance production occurred in the same place at the same time.

Changing one gene at a time, by non-coordinated step-wise mutations, cannot produce an increase in complexity, or a new organ, or for that matter produce a new species. Multiple simultaneous mutations on multiple genes are an absolute requirement. But worse than requiring chance evolution to provide simultaneous mutations on multiple genes, we must also require chance mutation to simultaneously build male and female counterparts, with built-in mating instincts.

The very concept of neo-Darwinism now becomes utterly ridiculous, that chance mutation in a population could produce male and female versions of the same species mutation, at the same place on earth at the same time, with the proper mating instincts, and proper matching sexual organs.

Until evolutionists can demonstrate how such a phenomenon could have occurred, using the empirical methods of science, this instance of irreducible complexity should constitute a major part of nailing-the-coffin-shut for Darwinism.

The point to be made is simple: sexually reproducing organisms are specifically designed to invest part of their energy producing mechanisms in the production of gametes and more individuals of their own species. Males and females, together, are one system.

Whether the concept of "cumulative selection" or not, we are still forced to ask what happened to the initial prototypes in which the systems were not yet complete, those that should still be undergoing their stepwise integration? A partially complete system could not have worked.

Thomas Colville, program director of the veterinary technology program at North Dakota State University, and Joanna Bassert, professor and director of the program of veterinary technology at Manor College in Jenkintown Pennsylvania, introduced the topic of sexual reproduction this way:

"The reproductive system is an oddity. The rest of the body's systems work to ensure the survival of the animal they are part of, whereas the reproductive system works to ensure the survival of the *species* of an animal. It interacts with other body systems, but purely reproductive structures are not essential to the life of the animal." Think about it.

"Another difference from other body systems is that the reproductive system requires a second animal (of the opposite sex) to fully carry out its function, which is to produce a brand new animal (the offspring). What we refer to as the 'reproductive system' in an individual animal is actually only half a system. A complete reproductive system is made up of all the male reproductive organs and structures in one animal and all the female reproductive organs and structures in another. Both are necessary for an offspring to be produced."

If evolution were true, where was the "reproductive advantage" for the already-successful asexually reproducing organisms? Why would non-goal-directed selection favor organisms that were expending part of their energy reserves in producing a new type of reproduction, rather than using the same energy requirements to produce more of their own kind? Wouldn't that have been the greater reproductive success demanded by evolutionists as the test for the success of a species? Yes, it would.

Reptiles lay eggs. Hominids do not. Evolutionists might try to infer a stepwise development of mammals and Homo sapiens by referring to the egg-laying characteristic of monotremes, and the marsupial pouch (marsupium) of marsupials, as intervals in a stepwise process. Even so, the jump from reptiles to hominids is tremendous. But, the method of reproduction remains the same for both – sexual.

The point herein is not the concept of one species transitioning into another; the point we are now discussing is that it takes two types of animals representing one type of organism, i.e. males and females, to complete the reproductive system in a sexually reproducing species. And, under any circumstances, whether the organisms under consideration are reptiles, mammals, or any other type, the question remains the same. By what pathway did the first sexually reproducing organisms ascend from their asexually producing ancestors? In what pathway were the structures developed which would allow the two sets of alleles to be joined together again within the body of the female organism as a zygote, with the destiny of becoming a single new individual representative of the species, either male or female?

With regard to the sexual production of a zygote, the actual system by which this is done is ingenious. Male gametes are inserted directly into a protective invagination of the body of the female. But this method has a problem. How does one insert a flaccid penis into a vagina? Some lower mammals have a penile bone. But, human males do not. So, how does one go about producing an erection firm enough for vaginal insertion?

Erections are produced by the release of nitric oxide, an increase in arterial pressure to the penis, and the closing down of the veins leading away from the penis; thus, the penis is pneumatically inflated to its erect and firm condition. But there's more to the process: the penis also has to have the necessary syncytium of vascular channels capable of undergoing hydraulic engorgement – the corpus cavernosum provides the ability (the spongy tissue which composes a large part of the penis). Moreover, there has to be a method of producing a zygote ejaculate, and a method of ejaculating the semen so produced into the vagina. The process of producing sperm (male zygotes) and delivering them into the female vagina is not simple.

To begin with, spermatogenesis occurs in the seminiferous tubules of the testes, in the scrotum. Once produced, they are transported to and stored in the epididymis in large numbers, numbering in the hundreds of millions. After insertion into the vagina, frictional stimulation of the hyper-sensitive glans portion of the penis increases male arousal. The arousal causes the testes to be pulled upward. Nervous reflexes produce waves of contraction in the smooth muscle of the walls of the epididymis, causing the spermatozoa to move into the vas deferens, and thence into the urethra, along with seminal fluid produced by the seminal glands.

Once in the prostatic portion of the urethra, a sphincter in the prostate gland closes off the bladder and another closes off the penile urethra. The area between the two sphincters swells and fills with semen – like a syringe. Then, at the height of climax, the sphincter at the base of the penile urethra opens, and waves of contraction of the bulbocavernosa and ischiocavernosa musculature, in combination with rhythmic muscular contractions of the vas deferens, the seminal vesicles, and the prostate, force the semen out of the male urethra and into the female vagina.

Once these rhythmic contractions have begun, they cannot be stopped until ejaculation is complete. This process would have been required from the time of the first prototypes.

The female is not passive during coitus. With arousal, vaginal lubrication is produced. Vaginal secretions lubricate the vagina and facilitate insertion of the penis without discomfort. In addition, the vagina expands and the cervix and uterus are lifted upward and away from the vagina, and the clitoris, labia majora, and the labia minora increase in size. And, under the proper conditions, the human female is also rewarded with the pleasant physiologic results of an orgasm.

It is obvious that the sexes were programmed, in advance, for the process of insemination. Can anyone possibly conceive of a chance stepwise process in which chance mutation and selection produced testicles, and epididymis, a vas deferens, a prostatic sphincter to shut off the bladder and prevent the flow of urine during ejaculation, a second sphincter at the base of the penis to capture an aliquot of ejaculate, and the integration of muscular activity to inject the aliquot into the female vagina, and make the stepwise process functional enough to produce insemination and fertilization during the development phase of the prototypes? The author can't.

As if the system wasn't already complicated enough, chance mutation and selection would had to have wired the whole system in such a way that there was an instinctive desire for copulation during the period of female heat in lower animals, and both the desire for and psychological reward of orgasmic ecstasy in response to copulation, as experienced by humans. In addition, as we shall read below, copulation in lower mammals occurs only when the proper hormones are circulating, the uterus is prepared for pregnancy, and the vagina is prepared for the trauma of coitus. How does the reader think chance mutation and selection might have gone about anticipating and integrating the proper psychological and physiological sex-related reactions into an overall functional mix – using only non-goal-directed chance mutations?

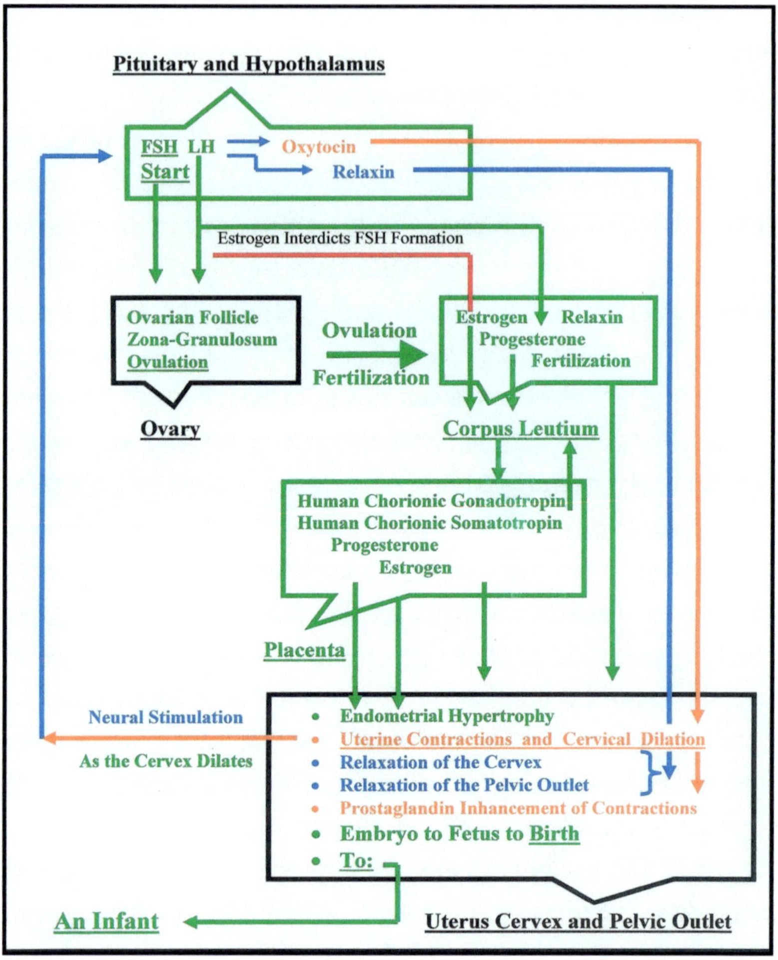

Simplified Schematic: Showing the Irreducible Complexity of a mammalian pregnancy and parturition. Not one element can be missing without causing failure of the pregnancy or dystosia.

It is impossible for a logically oriented mind, especially one trained in the concept of probability, to consider the process of sexual reproduction to have been engineered by anything other than design and irreducible complexity. If a single element of ovulation, fertilization and pregnancy are not in place, reproduction will fail because of polyspermy, failure of capacitation, too few sperm, or abnormal sperm, failure of implantation, or incongruous hormone production, i.e. birth control via "the pill."

Sexual reproduction is not amenable to Darwin's concept of stepwise evolution. The phenomenon of sexual reproduction can be explained only in terms of irreducible complexity and intelligent design.

ANOTHER EXAMPLE of a future goal-directed requirement for proper fertilization (impossible for random chance or natural selection to anticipate) is the fact that the heads of the spermatozoa carry a membrane enclosed sack, the acrosome, which contains a specific, antigen-specific, digestive enzyme capable of digesting its way through the zona-pellucidum of the specific species of ovum for which it is intended to unite, and only that specific type ovum.

This requirement had to have been present from the beginning. Who among evolutionists is willing to step forward and explain how the requirement was put in place from the beginning, before reproduction and natural selection had any "knowledge" of the requirement?

To prevent the waste of their digestive enzymes prior to reaching the ovum, the heads of the spermatozoa are covered with a coating that must be removed in order to activate the acrosome. This is done in the female reproductive tract by a process known as capacitation, which takes about four hours in humans, "the glycoproteins that make up the ovum's zona-pellucidum induce the sperm to undergo the acrosome reaction." Thus, inducing capacitation.

Those spermatozoa that are successful in reaching the ovum encounter the zona pellucida protecting the ovum and its haploid number of chromosomes. Each and every one of them try to digest their way through the zona pellucida, using the acrosomes with which they were provided at the time of their maturation in the male testis – unaware of the need for this specific requirement awaiting their arrival and "marriage" to a similarly unsuspecting, but similarly prepared, ovum.

The first sperm cell to penetrate the zona pellucida and deposit its haploid contingent of chromosomes to the formation of the diploid genome number provokes an instantaneous change in the zona pellucida, and within one second, the entire layer becomes virtually impenetrable by any other spermatozoan.

Moreover, the process of fertilization, as stated above, is species specific. The sperm of one species cannot penetrate the zona pellucida of another species.

The specific binding between the proteins of the sperm and the egg insures that sperm of other species cannot fertilize the egg.

FROM THE STANDPOINT of neo-Darwinism, any given asexual organism, which had never before reproduced sexually, would have had to have produced two separate, but sexually linked, types of organisms as a single reproductive unit, in a stepwise fashion, utilizing random mutation and natural selection and reproductive success. But, 99.9% of mutations are deleterious.

Moreover, what would there have been to select from prior to the first successful sexually-reproducing progeny? Wouldn't it be asking a great deal of probability to assume that the first sexually differentiated prototypes would have been successful at reproduction on the first try? Wouldn't evolution, randomly, have had to produce two separate organisms, a male and female type, in the same place at the same time on the first try? Wouldn't there have had to have been built-in programming in order for the female to ovulate during the same time interval that the male was ejaculating on the first try? And what would have happened to the entire experiment if the first stepwise trial was incorrect, in any aspect of fertilization, for any species on the first try?

Many ways to induce mutations are known, but none lead to new organisms. Mutation accumulation does not lead to new species or even to new organs or new tissues. If the egg and a batch of sperm of a mammal are subjected to mutation, yes, hereditary changes occur, but as was pointed out very early by Hermann J. Mueller (1890-1997), the Nobel Prize winner, who showed X-rays to be mutagenic in fruit flies, and remember, 99.9% of the mutations are deleterious.

Even professional evolutionary biologists are hard put to find mutations, experimentally induced or spontaneous, that lead in a positive method of evolutionary change.

The sexual reproductive mechanism is magnificent, but the concept of stepwise non-goal-directed mutations, in combination with non-goal-directed natural-selection in producing such a system, is unattainable.

Chapter Eight
Birth

Let's skip ahead and ask how the maternal body knows when the developing fetus is large enough and strong enough to sustain life in the outside world, but small enough to be delivered through the bony constraints of the female pelvic outlet? By what mechanism is the fetus extruded into the outside world? How does the mother keep from bleeding to death when the placenta tears loose from the uterus?

The answer, of course, is that every step had to be pre-programmed.

AFTER DELIVERY, the newborn infant is both defenseless and helpless. The lungs have never dealt with a gaseous atmosphere. The heart must switch from a three-chambered organ to a four-chambered heart, and pump blood through the newly aerated lungs, as well as the rest of the body. The immune system must be activated, as well as nourishment provided for continued cellular metabolism. The infant has only its built-in sucking reflex to work with. But, everything has been anticipated and pre-prepared by the time of birth – even the sucking reflex itself – and had to have been in place even for the very first prototype, in order for the initial mammalian prototype to survive.

The *foramin ovale,* which allows shunting of blood from the right atrium to the left atrium of the fetal heart (thus allowing 90% of the fetal circulation to escape the resistance to pulmonary blood flow), closes off. In addition, the *ductus arteriosus,* which carried the right ventricular blood that bypassed the lungs, clamps down – and the fetal circulation is converted to the four-chambered circulation that is necessary for optimal adult life. (Rhoads and Pflanzer p. 671; and Colville and Bassert p. 168-170).

In response to rising CO_2, and falling pH and O_2 levels, the not-yet-breathing infant's respiratory center is stimulated, the muscles of respiration expand the thorax, the lungs and their alveoli expand with ease, and the infant takes its first breath, contrary to laws of physics. The laws of physics demand that the infant's first breath produce a gas-liquid boundary layer in each alveolus with so much surface tension that drawing the first breath would be difficult in the extreme.

Though it is true that the newly formed gas-liquid boundary layers in the alvolae should produce tremendous surface tension, irreducible complexity has already anticipated the problem with production and storage of a special wetting agent in Type II pneumocytes, surfactant, which neutralizes the problem, even as it occurs. Type II pneumocytes produce and store surfactant only late in pregnancy, as the date of air-breathing draws near.

Death can occur in premature infants that are not old enough to have produced the special surfactants that neutralize pulmonary surface tension, because their muscles of respiration simply fatigue. How could non-goal-directed evolution have anticipated the need for a surfactant in the lung, especially so specialized surfactant as is present?

WITH REGARD TO THE INFANT'S NUTRITION, the mother's breasts have been preparing to respond to the infant's sucking reflex for months. Increased estrogen during pregnancy will have already stimulated the pituitary to release prolactin. The combination of estrogen, progesterone, chorionic somatotropinon and prolactin will have induced the mother's breast tissue to prepare for milk production – up to 1.5 liters per day. But, the hormone levels of the hormones noted above have to be balanced precisely, along with levels of thyroid stimulating hormone and adrenocorticotropic hormone, in order to produce milk. Why? Because if the hormone levels were not precisely balanced, mammalian mothers would produce needless quantities of wasted milk and energy throughout pregnancy.

Not surprisingly, it is estrogen and progesterone that actually block the production of milk during pregnancy. Thus, the same hormones that prepare the endometrium of the uterus for implantation and growth of the fetus also block the needless production of milk until the time of parturition, when the abandonment of the placenta results in a fall in estrogen and progesterone levels, coinciding with the newborn infant's need for nourishment and milk production.

However, the infant needs more than milk - a defense against disease is also required. Thus, the first material produced by the breast is not milk, it is colostrum. Colostrum confers the mother's immunity on the infant for about six months, if the colostrum is ingested early. Colostrum taken in too late in the neonatal period is simply treated as food, and digested, rather than allowed to confer immunity. So, colostrum has to be produced first, rather than milk, and ingested early.

With regard to milk production, per se, prolactin levels increase during pregnancy and reach high concentrations at birth. Following birth, prolactin levels diminish, and breasts respond more readily to suckling.

At that point, it is the infant's nursing that stimulates the hypothalamus of the mother by neural pathways which can produce a tenfold increase in prolactin production and milk formation.

Much of this process of inception and birth, we observe without deep thought as to the complexity of the amazing conditions that have to be in place for the common birth of a child. It has to be by design and not chance.

Chapter Nine
The Eye

One of the most intriguing aspects of any species is the eye. It is perhaps the most complex organ of the body, and not easily explained as something that happened by chance.

Anyone with practical experience knows the complexity of modern-day cameras. No one believes that such instruments come into existence by chance. So, why would anyone believe that human eyes, which are far more complex than cameras, occurred by chance? Evolutionists take a tangential approach to answering the question.

They invoke the fairy-tale concept of mutation and selection without ever explaining or empirically demonstrating any of the specific pathways, by which light-sensitive eye-spots might have developed in the first place, or transitioned into the various types of eyes used by the various animals of the animal kingdom. In the second place, we are asked to believe not only that such development in transition might have occurred, but that it did occur, and take the whole concept on faith.

It is because the entire complexity of the form and function of the eye is so difficult to grasp, that evolutionists are able to sidestep the demands of the scientific method with their semantically plausible hyperbole.

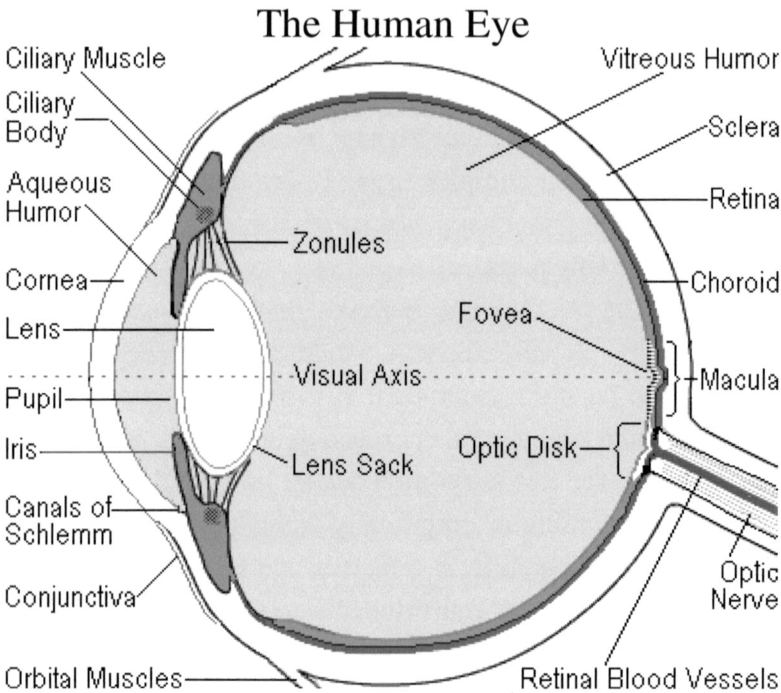

Cross section drawing of the eye (side view) with major parts labeled.

There are several evolutionary questions about the development and function of the eye, which evolutionists cannot dodge:

By what pathway could chance have produced and integrated color vision, scotopic vision, and photopic vision?

How could organisms migrating from a water habitat transition to dry land while maintaining vision without simultaneously changing the refractive index of their eyes?

How could non-goal directed evolution have apportioned the correct usable mutations to the genomes of the correct developing species in a successful progression?

How could lateral vision have transitioned to binocular vision while maintaining functional, usable vision during the process?

In general, moving lateral-facing eyes to forward-facing eyes is difficult in itself. The move requires a change in the design and physiognomy of the skull, and rearrangement of the neural innervation of the eye, and production of new integrated circuits. But more than that, a negative feedback system has to be developed to force both forward-facing eyes to focus on the same fixation-point.

In addition, spatially different portions of the two retinas must be made to display the same integrated information on the visual interpreting portions of the optic lobes, or the visual corticies, of the brain.

Thus, there must, perforce, be simultaneously integrated changes on various genes, including those that control the physiognomy of the skull, and the pathways of the nerves and blood vessels that supply the eye. This, in addition to the transposition of the muscles required for ocular control and movements of the eye globe itself, as well as the neural mechanisms which allow point fixation and cortical interpretation of visual images.

All transitioning occurring by chance, on various genes, simultaneously. Moreover, such changes during the transition process must occur in such a way as to allow an organism to remain functional and survive.

Intuitively, we know that such a transition is extremely improbable, to the point of vanishing probability. But unfortunately, we do not know enough about the undisclosed methods of transitioning from lateral facing eyes to forward facing eyes to apply the mathematics of probability to prove or disprove the concept with respect to the integration of function of the motor cortex, the basal ganglia, and the cerebellum of the brain.

We can discuss the physics and physiology of sight, lay the parameters on the table, and see if it is conceivable that the multiple integrated combinations of factors required for sight and binocular vision might have occurred by chance.

The chance development of the chemical basis for vision is improbable in the extreme, but rather than become entangled in the biochemistry of vision concerning rhodopsin, photopsin, or the bleaching and restoration of pigments, and so forth, we will, instead, confine our discussion of sight to the physics and biophysics of vision, which are more easily understood as indicators of the improbability of chance origin of sight itself, and binocular vision in particular. It is by discussing the improbability of chance that we prove the probability of design and irreducible complexity, in the production of both sight and binocular vision.

LIGHT IS A TYPE OF ELECTROMAGNETIC field, composed of two inseparable components: electric and magnetic portions. Electromagnetic fields can be in a relative steady-state such as a North-South orientation of a magnet, or the 60 cycle oscillating current of power lines, or they can oscillate in radio frequencies, microwave frequencies, infrared (non-ionizing) frequencies, or at the frequencies of x-rays or gamma radiation (ionizing frequencies).

When we humans think of light, we think of a specific portion of the electromagnetic spectrum that activates our human visual systems. Light to us is a specific portion of the electromagnetic spectrum ranging from wavelengths of about 400 nanometers (blue light) to approximately 750 nanometers (red light). This is the range of quantified energy to which our electromagnetic sensors respond.

Each wavelength carries a specific amount of energy that is quantified as photons. Photons in the blue portion of the spectrum carry more energy than those in the red portion of the spectrum, but any of the photons in between can activate our visual receptor mechanisms; in fact, receivers can be made to react to almost any frequency of electromagnetic radiation. In this respect, we have night vision goggles, infrared sensors, microwave receivers, radio receivers, x-ray badges for x-ray technicians, Geiger counters, etc.

Moreover, we can have wide-angle receivers, narrow-angle receivers, and even waveguides as design characteristics for various receptors. In addition, electromagnetic receivers can be modified with regard to brightness and contrast, and as anyone who has dealt with the first generations of television receivers will remember, we were frequently adjusting brightness and contrast in our black-and-white television sets.

Thus, it is a fact that understandable visual-information can be transmitted via electromagnetic means by simply employing appropriate shades of gray in combination with brightness and contrast. Color vision is nice, but it is not essential for the transmission of understandable information about a visual panorama.

So, who needs color vision to grasp the meaning or syntax of something that can be visualized in black-and-white? Why would any organism need anything more than black-and-white vision? Why would an organism need anything more than simple receivers, anything more than receivers capable of discerning the presence or absence of electromagnetic radiation in a "pixelated" visual field?

If organisms were already functioning satisfactorily with black-and-white vision (as do dogs), then how could non-goal-directed natural selection pick out mutations necessary to build a second visual process? Certainly, the organisms trying to use any new, developing method would be at a disadvantage - either wasted-energy-wise or vision-wise until the new method was perfected.

Natural selection would demand that such organisms undergoing such wasted, non-goal-directed visual development be eliminated. And it would make no difference which came first, black-and-white vision or color vision.

Accordingly, if black-and-white vision is adequate, is there even a reason for color vision, or vice versa? Well, there is a reason. Black-and-white allows usable vision, under low-light conditions, and color vision allows "color coding," under bright light conditions, of the things we see, which imparts instant information. On a black-and-white television receiver, we can see the outline of an orange or a red-delicious apple, or a banana; but can we tell if they are ripe or near ripe? The answer is no. Yet, on a color television screen, or in our environment, we know the answer instantly.

OK, how can one produce a sensation of color from a continuous band of electromagnetic radiation? We, as humans, already knew about the trichromacy theory of color perception when color television was produced, but how did "nature," starting from scratch, become aware that electromagnetic fields even existed, or determine how to use such electromagnetic fields to produce color by chance?

It is easy to understand the production of a visual message using the presence or absence of light, or changes in the luminescence/brightness of light, or contrasting shades of gray; but how does one produce color, or more importantly, a method of experiencing the sensation of color?

Well, "nature" and industry both used the same method, i.e. the trichromacy interaction theory. Industry broadcasts television signals in color by sensing electromagnetic information simultaneously in the three primary color ranges, and then transmitting three corresponding signals.

Television receivers react to those signals and redisplay each of the pictures created by each of the primary color cameras on our television screens, within the "critical reception duration" (photon collection interval) of each of our three types of color detecting retinal receivers (the three types of cones), and merge them into a combined color picture as interpreted by our optic cortices. And, of course, the same thing happens as we visualize color transmitted directly to eyes from the environment around us in nature.

So, the physical transmission, reception and integration of various overlapping wavelengths to form a color picture is understandable, but what is the actual physiological mechanism that converts a continuous spectrum of light radiating from our television screens, or our environments, into the sensation of color?

Our three types of cone-receptors, tuned separately to red, green and blue, do the same thing as the television receptor/cameras at the television studio that transduce light input into electronic output. The cone receptors in our retinas respond to the same variation in intensity and duration of signal in each of the three primary color ranges as do the three types of television camera sensors. The varying degrees of overlap in the stimulation of the three types of cone-receptors produce varying degrees of the overlapping sensations of red, green and blue, thus producing color in the same fashion that overlapping mixtures of food produce varying flavors.

In addition to the types of cone-receptors described above, some birds sense color by using colored droplets of oil in four or five types of cone-receptors.

Diurnal birds usually have yellow, red, green, and orange droplets. However, we should note that nocturnal birds have pale or transparent droplets. In the absence of adequate light, it is difficult or impossible to detect color anyhow. So, design considerations would reveal that color sensors would be wasted on nocturnal birds, thus, no specialized color receptors for nocturnal birds.

Does anyone who understands how difficult it was to develop color television believe that perception of color by animals came about by chance, or that diurnal birds possess a special ability to perceive color, or that nocturnal bird's lack of such color perception is by accident? Hardly!

TO THIS POINT, WE HAVE discussed two types of "seeing," i.e.: in shades of gray or in the mode we call color. But there are yet others. Some insects can see in the ultraviolet range. This is what helps honeybees to locate certain nectar bearing flowers, and which helps certain birds to recognize ripe fruit. Another type of sight is the kind specifically designed for photo enhancement in low-light conditions, as well as the perception of slight amounts of motion in a visual field.

Insects that require nothing more than a general idea of what they are looking at, but need to know about color and the slightest amount of motion in their environments, have compound eyes made up of ommitidia, or wide-angle waveguides containing multiple photo receptors.

Each of the multiple ommitidia points is angled in slightly different directions to cover their visual fields. Thus, changes in any given group of ommitidiae's fields of "vision" is readily perceived and transmitted to the brain as motion. Ommitidiae are both excellent color detectors, as well as motion detectors.

However, there is a drawback to this system: the overlapping wide angles of incidence of the waveguides produce a fuzzy image for the brain to interpret. On the other hand, why would the types of animals that possess compound eyes need anything more than the input of fuzzy outlines, the detection of color, and evidence of motion in their environments in order to function successfully? The answer is nothing; and, indeed, there are those that do function successfully, sometimes too successfully, especially flies when one is trying to swat them.

But, what does one do when one needs precise visual information? How could a jumping spider function with ommitidia type eyes? How could it visualize its prey well enough to catch it? Well, to improve the acuity of vision, one uses a lens and focuses the available incident light on as sensitive a group of individually queried photoreceptors as one can produce, with as small a number as possible cross-linked to a second-order bipolar collecting neuron.

Jumping spiders have eyes with lenses, and so do all animals that search for food in the presence of light. Birds of prey in particular need sharp vision. In nature, many species of birds have cones that are each individually wired to a bipolar transmission neuron, thus acting as individual pixels of a television screen or computer monitor.

Jumping Wolf Spider's Eight Eyes

The greater the number of pixels/receptors/color-cones in the fovea of a retina (the area where light is primarily focused), the greater the definition of the images we see.

Some diurnal birds, such as hawks, parrots and hummingbirds, have a second, temporally placed fovea; and some predatory felines, such as cheetahs, have a visual strip across their retinas which correspond with their horizon of vision.

Is there anyone who can really believe that perfect matching of form and function came about by chance mutation on various genes individually responsible for various parts of an animal's physiognomy and coalesced in perfect harmony by chance?

The concept of natural selection, although spurious, could understandably have selected the brightest and best for "reproductive success," but selection could not have sorted and distributed the correct optical requirements, tailored to coincide with the individual needs of the form and function, of the various manifestations of the spectrum of animal life. Sorting and proper distribution can only be done by design.

WOULD ONE TYPE OF VISION suffice for all vertebrates? No; just as insects, such as jumping spiders, need lenses rather than ommitidia, so vertebrates need vision that is matched to the conditions under which they specifically function. Evolutionists will immediately respond that natural selection does the sorting. But selection can only select from what is served up by "mutations."

This is not a proper method of distribution because mutations can go either way: for good or for bad. And, as everyone knows, mutations are overwhelmingly deleterious. So why would diurnal animals end up with the type of vision that is perfectly matched to their needs; and why would nocturnal animals possess the perfectly matched type of vision that they need?

One could understand the correct matching in some species, but not in all of them. When chance mutation is employed as a means of a genome modification, probability demands some mismatches. The mutation and selection route is so overtly improbable that it is impossible to imagine there is anyone who could believe it possible, except among those who believe in the magical, but spurious, concept of cumulative selection. To the logically functioning scientific mind, vision is a product of irreducible complexity, specifically designed to meet the needs of the form and function of a given organism.

Let's take ourselves, for example. As humans, and diurnal animals, our retinas are capable of producing two types of visual displays on our visual cortices. One is the type of vision that occurs under low-light conditions, so-called scotopic vision produced by the rod-type of photon receptors in our retinas.

The other type of vision is that which occurs in the presence of abundant light, that is photopic vision, namely the type of color vision produced by the cone-type of photon sensors in our retinas – especially those concentrated in the fovea centralis, the area where incident nodal rays are primarily directed.

The entire eye is amazingly complex, however, scotopic and photopic vision are unusually complex. The rod and cone receptors in our retinas are constructed and wired so as to be both spatial and temporal photomultipliers. For the most part, rods are cross-linked in such a way that the combined signals from a group of rods can, spatially, produce an increase signal capable of transcending the threshold for signal perception, by summation of signal.

This spatial method of collecting photons is further enhanced by a duration of excitation method that allows a given rod to individually collect several photons, over a period of time (temporal photo multiplication), to produce summated excitation and, thus, send its message to the spatial photomultiplier.

This, again, is a sophisticated method of transcending the threshold for incident photon perception. In addition, light collection is further enhanced by opening the lens aperture, iris opening, of the eye, and also allowing the eye to become dark adapted by the production of increased amounts of light-sensitive pigments, in particular rhodopsin. What might be the sensitivity of this system? The amazing answer is one photon per rod-receptor will stimulate a receptor.

BEFORE PROCEEDING TO A DISCUSSION of why the transformation of lateral vision to binocular vision is impossible, let's first perform a bit of housekeeping and discuss some points of particular interest for those who are technically oriented. Firstly, our eyes automatically converge their optic axes onto the same spot of interest in our visual fields as a "fixation point" upon which to focus our gaze.

The focus point of interest is focused on corresponding fields of vision in the corresponding parts of our retinas – and with the nodal beams concentrated on the fovea centralis of each eye – again, the area where our cones occur in the highest concentration, and thus where it is possible to detect the greatest detail regarding an object of interest in our visual field.

Secondly, our eyes blink about 15 times every minute, automatically, keeping our eyes moist, washing away debris, and providing protection from bacteria by the presence of lysozyme in our tears. But, even when blinking, we still keep the object we are looking at in the same fixation point and always in focus when we again open our eyes. (Try it. Close your eyes, then open them; whatever you were looking at will still be in focus.)

This is not true in industrially produced optically-controlled feedback instruments. In industry, loss of signal, whether due to changes in the noise to signal ratio or the complete loss of signal for any reason, can result in temporary loss of information tracking. In animal life, this could be a catastrophe for an organism.

Receptor packing minimizes the problem of signal loss, but, if the packing is too tight, "cross-talk" results. This results in diminished energy transmission, and a reduction in sensitivity. This so-called "frustrated total reflection" due to too tight packing does not occur if fibers are separated by one micron.

Not surprisingly, retinal fibers in some animals are, indeed, packed to the limit of about one micron. (cones are about 1.5 to 2 times more tightly packed in birds than in other vertebrates). "Evolution" appears to be efficient, indeed!

Really?

Well, maybe not. Let's digress a little and talk more about the eyes of birds, and the fish from which they, and we, theoretically, "evolved."

Birds fly through trees, brush and dust. Some dive into oceans, lakes or pools of water to obtain vegetation or prey. Birds need to protect their eyes. They have three eyelids: an upper lid, a lower lid, and a nictitating membrane.

The nictitating membrane is transparent and moisturizes the eye. Some species of diving birds have nictitating membranes with a clear area in the center, which acts as an underwater contact lens. But, is the reader aware that fish have no eyelids? They don't need to moisten their corneas, and they can hide in dark places in pools or streams or in the ocean depths, in order to protect themselves from the sun or predators.

So what happened to the first fish that came out of the water onto land? The answer is simple: they were blinded by the glare of the sun. Their corneas dried out, and they either went back into the water, or they went blind. Even if they were able to see when out of the water, with a refractive index designed for water and not for air, what would they have found to eat?

It would be speculative to believe they could have captured insects. And, even if there were foliage in the area, it could have been poisonous. We will discuss mammalian lungs and the transition from amphibians to land dwelling vertebrates in the next section.

It is impossible to believe that organisms with such poorly prepared eyes could have come out of the water and lived long enough to develop eyelids, lacrimal glands, tear ducts, corneas and lens with the proper refractive index for air, and also transition lateral facing eyes into forward facing eyes and binocular vision. There are many other reasons that fish, such as the lung fish, could not have transitioned from water to land, i.e. from fish to amphibians. (A second book on the subject, titled *Intelligent Design of the Species*, has been written, which includes more information on the subject.)

If those who teach "cumulative selection" could have carried out such a transition, let them explain how. And, let's all remember that using the spurious concept of cumulative selection is only adding "once upon a time" to the fairy-tale under consideration.

In science, it is not enough to suggest that something might have happened. It is necessary to elucidate the pathway by which it did happen. The relatively new scientific discipline of evolutionary development is trying to find such evolutionary pathways. However, what is more likely to happen is that the science of evolutionary development will end up falsifying the concept of evolution for lack of evidence of reasonable pathways.

THE EVOLUTIONISTS ARE PEOPLE who may be very well versed in their own narrow fields of expertise, but do not really understand the intricacies of the broad field of science. Let's take, for example, the prattle of two well-known, but uninformed, biologists and authors:

Frank Zindler, a professor of biology and geology, stated the following: "as an organ developed via the opportunistic twists and turns of evolutionary process, the human eye is explainable. As an organ designed, the human eye is inexcusable. For unlike the invertebrate eyes, the human eye is constructed upon the foundation of an almost incredible error:

The retina has been put together backwards! Unlike the retinas of octopuses and squids, in which the light gathering cells are aimed forward, toward the source of incoming light, the photoreceptor cells (the so-called rods and cones) of the human retina are aimed backwards, away from the light source.

Worse yet, the nerve fibers which must carry signals from the retina to the brain must pass in front of the receptor cells, partially impeding the penetration of light to the receptors. Only a blasphemer would attribute such a situation to – design!"

Similarly, in his book *The Pony Fish's Glow*, biologist George Williams recorded the following: "Some eye features are not merely arbitrary but clearly dysfunctional. The nerve fibers from the retinal rods and cones extend not inward toward the brain but outward toward the chamber of the eye and source of light. They have to gather into a bundle, the optic nerve, inside the eye, and exit via a hole in the retina.

Even though the obstructing layer is microscopically thin, some light is lost from having to pass through the layer of nerve fibers and ganglia and especially the blood vessels that serve them. The eye is blind where the optic nerve exits through its hole. The loose application of the retina to the underlying sclera makes the eye vulnerable to the serious medical problem of detached retina. It would not be if the nerve fibers passed through the sclera and formed the optic nerve behind the eye. This functionally sensible arrangement is in fact what is found in the eye of a squid and other mollusks."

These comments by biologists Zindler and Williams have forever been recorded in the annals of history to their chagrin and discredit, in the eyes of those who know better.

Zindler and Williams failed to recognize the problem that exists for diurnal animals: namely that of too much light during the daytime, with the resultant problem of glare, and the problem of how to remove the debris that would break away from the cells of the retina and fall into the vitreous chamber, if the rods and cones were not turned backward into the choroid coat. It is the function of the choroid layer to carry away cellular debris, and, at the same time, cut down on excess photon stimulation, i.e. glare, a problem which we will discuss later.

As if one blunder were not enough, Williams made yet another in discussing the form and function of the human eye in the *The Pony Fish's Glow.* He stated the following: "Not all features of the human eye make functional sense. Some are arbitrary. To begin at the grossest level, is there a good functional reason for having two eyes? Why not one or three or some other number?

Yes, there is a reason: two is better than one because they permit stereoscopic vision in the gathering of three-dimensional information about the environment. But three would be better still.

We could have our stereoscopic view of what lies ahead plus another eye to warn us of what might be sneaking up behind. When we examine each eye from behind, we find that there are six tiny muscles that move it so that it can point in different directions. Why six? Properly spaced and coordinated, three would suffice, just as three is an adequate number of legs for a photographer's tripod. The paucity of eyes and their excess of muscles seem to have no functional explanation."

Williams' statement is ridiculous. Of course, there is a functional explanation. It is just that biologist Williams is unaware of the answer. Perhaps he should have done more reading and less writing.

Firstly, if one were to have used Williams' poorly thought out design, the globe of the eye would have been jammed into the back of the eye socket by the contraction of his three muscles, thus causing the globe of the eye to impinge against the back of the eye socket, thus making motions of the eyeball difficult.

Secondly, the posterior aspect of the eyeball would have been flattened posteriorly, thus the retina would have been distorted – making vision variable. Fortunately for us all, the muscles required for motions of the eye were designed for efficiency and balance, and to avoid the problem of flattening the posterior aspect of the eye against the eye socket of the skull. The superior oblique and inferior oblique muscles (dorsal oblique and ventral oblique in canines) counteract the pull of the rectus muscles so as to balance the eyeball within the eye socket during eye motion.

We are all indebted to this irreducibly complex design, since Williams' suggested design would not have worked. There must be counterbalanced dynamic forces in order to allow the eye to move properly (the same thing is true with motions of the head of the humerus within the glenoid of the scapula – with regard to shoulder motion and the prevention of jamming).

We present this example of irreducible complexity with the admonition that, like Dawkins' false concept of cumulative selection, evolutionists will attempt to dodge the implications of this conundrum by using their usual fall-backed position, which is to argue that the genetics of vision is a "highly conserved function" (whatever that might mean in the real world of empiric science), and that binocular vision "evolved" by a route different from that of lateral vision. What a distortion of science and the scientific method.

Chapter Ten

The Anatomy of the Trachea and Esophagus

Integration of respiration and swallowing is interesting in its function. The cough mechanism is an impressive design element of the respiratory system. However, let's move on to an even more innovative aspect of respiration – the integration of breathing and swallowing.

What would be the design considerations if the reader were designing a human body? Would you put the respiratory passages and the food swallowing mechanism close together? After consideration, the answer is absolutely yes. By combining the airway and the swallowing channel into a common oral pharynx and nasopharynx, the body is provided with alternative access to continued respiration, should the nasal passages become blocked by infection, trauma, or tumor, etc. It would be the rare ungrateful person who is not extremely pleased with this element of design, when a cold or the flu blocks nasal breathing, and forces one to breathe through one's mouth. Being able to open one's mouth and breathe freely is ecstasy for all those who have experienced the severe difficulty that arises when trying to breathe through a stuffy nose.

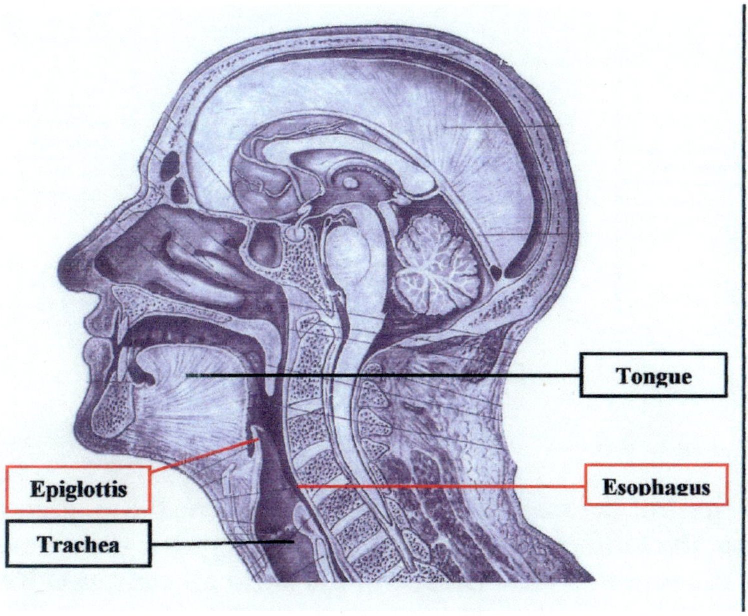

Mid-sagittal view of the mouth, nose, nasal-pharynx, traches, and esophagus. In addition, this diagram shows the tongue, base of the tongue, epiglottis, hard palate, soft palate, nasal trubinates, spine, brain, brain stem, and other soft tissue.

OK, are there any other advantages to having the airway and the swallowing mechanism close together? Again, the answer is yes. Taste is in large part a function of smell. Try plugging your nose and swallowing some tasty morsel. See if you can taste it. In many cases you can't. And, under any circumstances, only the most basic elements of taste can be appreciated when the nose is plugged, i.e. salty, sour, etc. However, more important than taste is the ability to detect the putrid odors that arises from things that are deleterious.

Lung Schematic – Level of the Tracheal Bifurcation

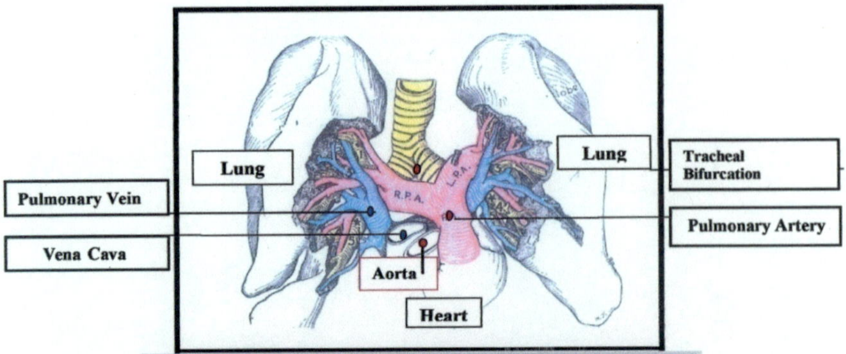

Lung Hilus: A single cardio-pulmonary unit. There is no room for an esophagus. The Aorta, Superior Vena Cava, nerves, lymph nodes, and pericardial sack have all been cut away to show the trachea.

In this part of the mediastinum, the trachea, the bifurcation of the trachea, the bifurcation of the pulmonary arteries, the pulmonary veins, the superior vena cava, and the aorta all converge. Embryologicly, the cardiopulmonary system develops as a unit. There is no room for an esophagus to tread its way through this area. If the esophagus were to lie in the anterior mediastinum, in front of the heart, it would have to make a sharp bend directed posteriorly to get back to its hiatus through the diaphragm (separated from the descending aorta by the diaphragmatic crurae) in order to enter the abdomen. It would be difficult for a bolus of food to negotiate the required sharp posterior turn between the heart and the diaphragm. The only logical location for the esophagus is in the posterior mediastinum, behind the trachea, and adjacent to the spinal column where it takes a straight course until turning slightly anterior, in passing between the crurae of the diaphragm. (Modified from Grant's Atlas of Anatomy 7th Edition.)

Chapter Eleven
Swallowing

Once we have decided that there are good reasons for having the airway and the swallowing mechanism in close association, what are the design problems we must face? Here are some major considerations:

1. How could we ask embryology to channel an esophagus through the maze of structures that coalesce in the midline of the mediastinum of the thorax? We already have a major tangle of major structures: such as the aorta, the superior vena cava, the pulmonary arteries, and the pulmonary veins, as well as an accumulation of lymph nodes and nerves – among which are the Vagus and Cardiac nerves.

2. Should the trachea be in front of the esophagus or behind the esophagus? If we put the esophagus in front of the trachea, we would have the advantage of food and fluids passing directly down the food-carrying channel, without having to cross the airway. But, if we did that, the esophagus would lie in front of the trachea and bronchi as it transverses the thoracic cavity. And, since the cardiopulmonary systems develop together, wouldn't such a design interfere with the anatomy and function of this choke point of immovable structures? The answer is yes.

Therefore, the only other available route into the abdomen would be through the anterior mediastinum. But there is a serious problem. The esophagus would have to first make a sharp bend, anteriorly, to move in front of the pericardial sack, and then, after traversing distalward, would have to bend sharply, posteriorly again, to find its way through the diaphragmatic muscle without injuring its structural and functional competence, as is true in the case of the safely engineered normal anatomy.

So, a primary design consideration is that the esophagus must lie behind the trachea and the cardiopulmonary system, in traversing the thorax. This can be arranged by running the esophagus through the thoracic cavity close to the spinal column in the retropleural space. This is what the body does.

3. If it is a primary design consideration that the esophagus lie behind the cardio-respiratory system, how can we keep food and water from entering the trachea as food and water enter the mouth, and cross the trachea, in order to reach the esophagus? Does the reader have any ideas? Do our evolutionary biologist friends have any ideas? Would random chance come up with the correct solution? The design of the body did. And, the design is excellent – as attested to by the fact that we humans can eat, drink, breathe, and talk as much as we want – with only a very rare episode of choking. And, if we do start to choke, we have the cough reflex to protect us.

How did design accomplish this amazing functional fete that allows food and liquids to pass across the airway without dropping inside? It was done using a combination of structural elements and nineteen pairs of small muscles, plus the superior and inferior sphincters of the esophagus, for a total of 41 muscles, all computer-controlled, in synchrony, by the automatic swallowing mechanism of the swallowing center of the brain.

The swallowing mechanism has been evaluated by cine-radiography, EMG, animal experiments, and the excellent work of Bosma, J. F. (Deglutition Pharyngeal cap stage; Physiological Preview No. 371957 pp. 275–300), Doty, R. (Neural Organization, section 6, Elementary Canal; volume 4; American Physiological Society,

Washington, pp. 1861–1902), and Hiiemae, K. M. (Mobility and Mastication – a review of the activity of the jaw muscles and the movements they produce in chewing; Development, Function, and Evolution of the Teeth, 1978; Academic Press, New York and London, pp. 359-398). The process of swallowing, or deglutition as it is properly called, is as follows:

VOLUNTARY FIRST STAGE OF swallowing is a unique mechanism. The swallowing mechanism is divided into three stages. The first part is under voluntary control – another excellent design feature. Why, because as complicated as the swallowing mechanism is, an automatic computerized method is essential. However, one would not want to start that automatic process just by putting something in one's mouth. If one were to find a fluid to be fetid, or food to be rancid, one would want to be able to spit either the food or the fluid out of one's mouth, without swallowing it. So, starting the swallowing process by voluntary action, rather than automatically, is a very fine design characteristic.

The first stage of swallowing is initiated by putting the tip of the tongue against the roof of the mouth – the hard palate, and pressing the rest of the tongue against the back part of the hard palate. This action separates a bolus of food from whatever else is in the mouth, and pushes the bolus backwards towards the rear of the throat. At that point, the swallowing mechanism takes over, and swallowing becomes automatic through what are known as the second and third stages of deglutition.

As food reaches the back of the throat, a set of muscles raise the soft, rear-portion of the roof of the mouth, the soft-palate, and squeeze it against the back of the throat, thus shutting off the nasopharynx, so that food cannot go up the nose (another excellent design feature). At the same time, several small muscles automatically pull the hyoid bone and the larynx upward, pressing the opening of the trachea against the base of the tongue and the epiglottis.

Therefore, this is partially occluding the mouth of the trachea. Moreover, breathing stops, the vocal cords close simultaneously, and the mouth of the trachea undergoes a "purse-string closure" as the aryepiglottic folds, that surround the entrance to the trachea, close on one another. The reader can check the lifting and forward thrusting of the larynx and the hyoid bone, which lies just superior to the larynx, for him or herself by placing the thumb and index fingers on each side of the larynx and swallowing. You will feel the throat move upward and forward. Also, try to breathe while you are swallowing. You will not be able to! You cannot, voluntarily, override the protective design characteristics of the swallowing mechanism.

At the same time, as the bolus of food moves downward toward the esophagus, the epiglottis, a catilaginous protrusion of tissue at the base of the tongue itself, folds down, over the entrance of the entire above-mentioned mechanism. The bolus of food slides to either side of the tracheal entrance, around the epiglottis, and into the pyriform fossae on either side of the larynx – directed there by the aryepiglottic folds, held erect and open by the backwards pull of the posterior crico-arytenoids muscles, and then into the waiting, already opened, proximal sphincter of the esophagus – thus bypassing the entrance to the trachea. The entire process takes less than one second. Do you think you could manage this voluntarily? Can anyone believe that random chance organized this elegant sophisticated system? The author cannot.

The probability against the 19 pairs of muscles, plus the esophagus and its two sphincters, involved in the swallowing mechanism, being in the proper place, and oriented properly, or that the swallowing mechanism originated by chance is one chance in (2 X 19) +3; or 41 factorial; or approximately one chance in 3.3^{49} – roughly, less than one chance in 33, 000 billion, billion, billion, billion (see chapter 5 for the mathematics of probability in the non-condensed book).

To put that probability into perspective, it should be noted that slightly more than 1.7^{17} seconds have passed by since the origin of the earth – if the age of the earth is taken at 5.5 billion years. This computation means that probability could have allowed 3.3^{32} chance combinations = 190,000 billion billion billion chance combinations occurring every second, somewhere on this planet, since the earth became organized – and there would still be the probability that, in all that time, and with all those chances, the swallowing mechanism would, still, not yet have developed as it has.

That is the probability against the swallowing mechanism having been developed by chance. What is more, the actual probability is even far less, because we have not factored in the negative probability surrounding the proper nervous innervation of the 41 muscles involved, nor the software programming required to activate the swallowing center of the brain.

So, why don't we factor in these additional considerations, limiting these considerations to the bare minimum: namely, that there must be at least 41 nerves to innervate each of the 41 muscles, and there must be at least 41 subprogram " addresses" within the swallowing center responsible for correctly, sequentially activating each muscle in proper order. If that is correct, and it is, it means that there must be three times as many, properly integrated, sub-components within our swallowing mechanism than was initially stated above. That means that the probability against the swallowing mechanism having been developed by chance is (3 X 41), factorial. That number is 1.2 X 10^{205}. Science accepts negative probability numbers with exponents of 200 or greater, as indicating absolute improbability! And, remember this is only one system within the human body. Also, remember that probability is not concerned with either the method of production or the time required in producing a given result. The probability that a given result will have occurred is dependent only upon the number of factors to be considered (according to the fourth law of probability).

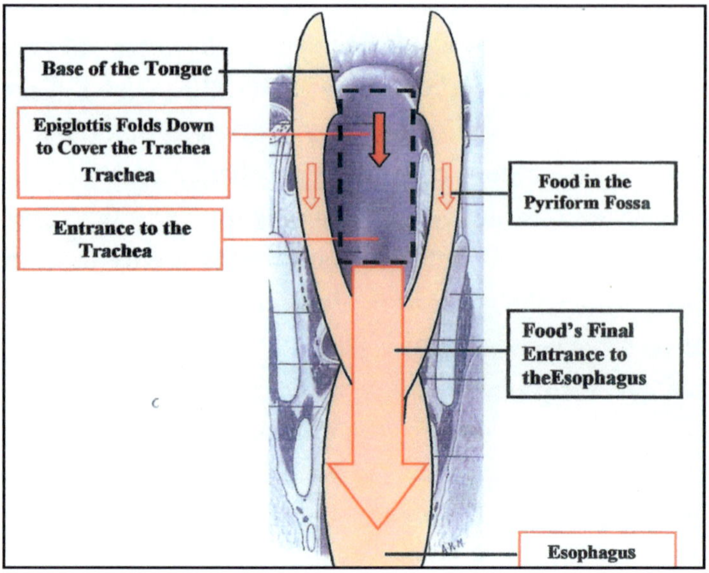

AUTOMATIC THIRD STAGE of swallowing has to do with not swallowing. When a person is not in the process of swallowing, the upper end of the esophagus located at the lower end of the throat is held tightly, and closed by a circular muscle – the proximal esophageal sphincter. This sphincter prevents air from entering the esophagus during breathing. Air in the esophagus could lead to air in the stomach and gut. And, anyone who has ever experienced "gas pains" knows that such air swallowing can lead to abdominal cramps, pain, and flatulence. Accordingly, we see two more excellent design features: a sphincter that keeps the top of the esophagus tightly closed and an automatic mechanism that relaxes that sphincter, and allows the esophagus to accept the bolus of food passed to it by the second stage of swallowing, described above.

As soon as the bolus of food enters the esophagus, the proximal sphincter closes, and a wave of relaxation proceeds downward along the esophagus, followed by a wave of contraction. The bolus of food is trapped between the two automatic waves of peristalsis. This pushes the bolus of food down the esophagus toward the stomach, where there is another sphincter which closes the stomach off from the lower end of the esophagus. The lower sphincter also opens and closes automatically, in sequence with the waves of contraction and relaxation. Thus, we encounter two more excellent design features.

The intra-thoracic pressure, the pressure within the thorax, is less than that in the abdomen. So, if there were no sphincter at the lower end of the esophagus to close the stomach off from the lower end of the esophagus, a person would develop painful erosions of the lower end of that passageway, as the stomach's digestive fluids refluxed back into the lower end of the esophagus, as a consequence of the differential pressure between the thorax and the abdomen, thus causing "heartburn." This condition can occur anyhow, in cases of hiatus hernia, where the hole in the diaphragm, through which the esophagus passes to get into the abdomen, is weakened or enlarged, and/or when the sphincter at the lower end of the esophagus is weak.

That is the value of the first of the two excellent design features to which we just alluded above. The second is that the sphincter at the lower end of the esophagus relaxes automatically, to receive the bolus of food destined for the stomach, allowing the bolus to pass through. The process, which moves food through the esophagus and into the stomach, takes about 10 seconds. It is called the primary wave of swallowing peristalsis. However, should a bolus of food become stuck in the esophagus, a secondary wave of swallowing peristalsis starts above the area where the bolus is stuck, and takes another try at pushing the bolus the rest of the way down the esophagus and into the stomach – amazing! Could this beautifully integrated process have been instituted by accident? Never!

Moreover, consider some of the other features of swallowing and breathing. Isn't it fortunate that we have parotid glands and salivary glands that pour fluid into our mouths, both to facilitate digestion (starches) and swallowing. Have you ever tried swallowing food when you are parched, or when you are dehydrated, on a hot day, and your mouth is dry? It is hard, to the point of being impossible. You have to work your tongue and jaws in a chewing motion to get enough saliva flowing so that you can swallow what you have in your mouth, either that or drink some type of fluid.

How fortunate that "evolution provided us with both a swallowing mechanism and a moisturizing system." But which came first, the moisturizing mechanism or the swallowing mechanism? Fish, from which we are postulated to have evolved, have plenty of surrounding moisture, and utilize ram type swallowing. They usually don't require any special mechanisms for swallowing, but what about the poor organisms that constituted the original life of land animals? How did they manage to swallow? How did they manage to chew dry-land food with their fish teeth? How did they manage to survive? As an aside, are you aware that our own chewing can be placed on automatic? It can be, fortunately, that allows us to think about other things while we are eating.

Lastly, isn't it fortunate that, if there were such a thing as evolution, random chance placed our salivary glands in our mouths, rather than in our ears, or our feet, or in our stomachs. And, isn't it fortunate that our olfactory nerve and its sensory end-organs of smell are located in the nose, rather than elsewhere?

And, isn't it fortunate that the nerves of smell pass directly through the cribriform plate of the skull, directly into the olfactory area of the brain, so that it takes only an instant to realize that something in front of our mouth doesn't smell right, or, if already in our mouth, doesn't taste right, giving us a chance to spit out whatever might be injurious, before being ingested and causing harm.

If, after considering the elegant design features that integrate and execute swallowing and breathing, together, there are still those who believe that all the muscles involved, and all the structural features, and all the coordination involved, could have occurred by chance, the author would appreciate an explanation. How could any single mutation, or any group of mutations, or any combination of mutations, instantly, or overtime, have occurred in such a manner in the correct position on the correct gene, on the proper chromosome, in the proper order, to allow the first organisms to function and to develop into the bodies we now utilize?

And, by what means were such "evolutionary changes" integrated into the embryology of already existing and functioning organisms, or those produced later during periods of increasing complexity, all without goals, without utilizing knowledge or intellect, and without employing engineering skills?

The very concept appears ridiculous to an educated mind. To believe differently is for evolutionary biologists to demonstrate, by factual experiment, how such amazing structural and functional integration could have been brought about by chance. And, please let them demonstrate how it's done, from start to finish, not just by speculation and postulating an unproven, never before observed mechanism resulting from the use of tortured logic. Rather, let them demonstrate a specific mechanism by starting from prokaryotic organisms, and continuing on to eukaryotic organisms, and finally to Homo-sapiens.

However, for the time being, we will be satisfied if evolutionists can simply design and implement a machine, such that its structure, function, software, and computerization can mimic that which they believe occurred by accident in the integration of our breathing and swallowing mechanisms. Again, don't tell us - show us! Oh, and by the way, evolutionists, be sure to keep liquids from leaking into the trachea during swallowing in the organisms you design; fits of coughing can be very troublesome.

AS WE HAVE OBSERVED FROM the discussion above, there are good reasons to have the respiratory and swallowing mechanisms closely integrated. If that is the case, then the swallowing mechanism and respiration must be integrated from the beginning. There cannot be slow, stepwise increments – as supposed by the concept of mutation and natural selection. As previously discussed, failure to protect the trachea and the lungs from foreign objects, such as food and fluids, perforce, can result in foreign body reactions and infection in the respiratory system. Thus, there is no alternative but to have the respiratory system protected from the very beginning. Otherwise, animals with the mammalian type respiratory systems would have a difficult time surviving.

If one were to try developing an integrated respiratory and swallowing mechanism on a stepwise basis, how would the reader, or anyone else, go about it? How would you protect the nasal cavity from food in the mouth being pushed into the nasopharynx and nose during swallowing?

How would you protect the trachea and the lungs? How would you move food from the mouth into the esophagus and into the stomach – in an integrated fashion, every time, without choking? And, how might proper mutations occur in the proper places at the proper time? And, how would random chance mutation go about properly integrating new elements of the swallowing mechanism into already-established, and therefore structurally fixed, and preexistent, components? Lastly, how would random chance go about "rewiring" the swallowing center and the involved muscles?

There can only be one answer. The integration of the respiratory and the swallowing mechanism cannot be done in a stepwise manner. It must be integrated all at once. There is no possible way in which the structures involved can be both sequentially produced, separated by variable periods of time, and then properly "rewired" or "rerouted." Moreover, during the period of time required to produce the new elements of the swallowing system, the new "wiring," and the "new routing," the involved animals would already have succumbed to the complications mentioned above.

Thus, if evolutionists want to continue believing in stepwise evolution by means of macromutations, "hopeful mutant monsters," neo-Darwinism, symbiogenesis, or by Kimura's theory of neutral mutations and random drift, the decision is theirs. They have the right to keep their faith. But, the onus remains on such believers to explain the scientific rationale for their ongoing beliefs. Until such time as evolutionary biologists, or population geneticists, can empirically demonstrate that stepwise construction and integration is a reality, design must be considered as the only possible method by which the integration of key parts of the body, such as the respiratory and swallowing mechanisms, were achieved.

Chapter Twelve
Avian and Mammalian Lungs

Just as we started with the biophysics of light in our discussion of the eye, we need to briefly discuss the concept of respiration. In general terms, respiration merely means an organism's intake of oxygen and output of carbon dioxide and water, plus heat. We have already called attention to the fact that form must be matched to function and circumstance in order for a complex system to work. Accordingly, as would be expected, respiratory mechanisms vary with the circumstances under which an organism must live.

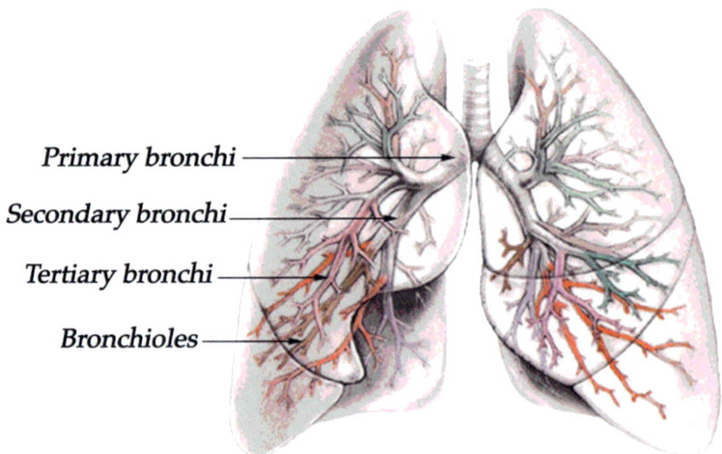

12.1 The Human Lungs

Some invertebrates use simple diffusion. Some insects use tracheal systems in which gas exchange occurs via small pores/spiricles in the outer covering of the organisms. These spiricles in turn lead into tubules that permeate the organism. Any given cell in a tracheal system is rarely more than three cells away from a tubule.

Another system is the gill system of fish. The system is one in which fish pump water over their gills by opening their mouths, closing the operculum which covers the outer aspect of the gills, lowering the floor of the buccal cavity, drawing in water, and then closing the mouth, opening the operculum, and raising the floor of the buccal cavity, which decreases its volume, and thus forces water into the opercular cavity, over the gills and out through the open operculum flap. As water passes over the thin-walled and delicate gills, the partial pressure gradient of oxygen in the ambient water forces oxygen through the gill walls and into the gill capillaries.

At the same time, the partial pressure gradient of carbon dioxide in the capillaries of the gills forces carbon dioxide out into the ambient water. In addition, the complexity is even greater than described above, in that the capillary flow of blood is opposite to the flow of the ambient water.

As the blood leaves the gill, it is in contact with the freshest water entering the gill chamber, giving the exiting blood an extra boost of oxygen. This countercurrent flow is important because the partial pressure of oxygen in water is far less than in the atmosphere. Moreover, the density and viscosity of water is far greater than that of air, accordingly the gills need all the help they can get in carrying out proper ventilation. And they do. The design is perfect for their habitat; again, an example of matching form and function to habitat. Since evolution is non-goal-directed, this perfect matching must, therefore, be evidence of design.

As regards air breathing, land dwelling organisms of the higher taxa, evolutionists would have us believe that they descended from either lungfish or coelacanths. The transition from water to land, currently, has the lungfish theory in vogue. The textbook Biology Concepts and Connections, published in 2003, describes it this way: "Historically, amphibians are of major importance because they were the first terrestrial vertebrates.

During the Devonian period, a diversity of plants and arthropods already inhabited the land. Plants rooted at the edges of the ponds and swamps created new living conditions and food supplies. Lungfish were already supplementing gills for gas exchange at water's edge, and sturdy muscular fins supported by extensions of the skeleton were probably better than fins for paddling or crawling through the dense vegetation. Thus, lungs and appendages evolved in certain fishes tens of millions of years before the earliest amphibians crawled on to land."

The hypothesis stated above is pure conjecture on the part of the authors. There is no way of verifying such a hypothesis. In fact, the empirical evidence contradicts the concept.

The gills of fish bear no resemblance to the respiratory mechanisms of amphibians or reptiles – the intermediates between fish and the ventilatory systems of birds and mammals, which are the topic of discussion in this section. The "lungs" of lungfish bear only a superficial resemblance to the ventilatory systems about which we will be talking. The lungs of lungfish are little more than air sacs with invaginations and blood vessels running through them.

It would appear that these lungs, at least in the case of the African lungfish, Protopterus, are designed for estivation – "summer sleep" – not for normal function. During periods of drought, when ponds dry out, lungfish survive by estivating in the drying mud at the bottom of ponds. They secrete a material which forms a cocoon-like dirt ball, with a tube leading from the mouth to the atmosphere above. Thusly, they can survive until the next rainy season. Is there any reader who believes that this just happened by accident? What happened to the first lungfish that didn't know how to make a dirt ball cocoon? Is this not another case of irreducible complexity?

Numerous authors, including Richard Dawkins, stated that if evolution were a matter of chance, it wouldn't work. We will discuss that point later. But, for the time being, let's just state that neo-Darwinism is a theory that is in just as much trouble as is Darwinism itself. Because someone or something has to serve up the proper combination of mutations before natural selection can take effect on the resultant phenotype.

Accordingly, you can have all the "cumulative selection" you want, but if the proper integrated combinations of "mutations" do not occur within a given population's genome, you cannot transition fish to lungfish, or lungfish to amphibians, or lungfish or amphibians to reptiles, or reptiles to birds or mammals, etc.

Let's consider the transition of fish to amphibians or reptiles, and let's start with the frog. Which came first - the adult frog or the tadpole? If the mutations in the precursor organism's genome were such that a tadpole was produced, then how, why, and where did the very complex mutations occur that provided a swimming tadpole with a tail and gills, and also with the wherewithal to produce lungs and legs, that would be required on land at a later date? Is this not a case of teleological mutation, i.e. goal-directed mutation?

Of course it is. How else could a set of mutations in an animal designed for a watery habitat have been provided, from the beginning, with the appendages required to live on land. For such a thing to have happened, mutations would have had to have occurred on control genes that would switch the expressive genes of fins into appendages with joints and the corresponding muscles for control, as well as proper eyes that would not dry out and would have the proper refractive index for air, rather than water.

In addition, there would had to have been a series of mutations capable of producing a latent conversion of gills to lung-type ventilation – with all the attendant changes in gas exchange and change in pH, etc.

Adult frogs breathe/ventilate their "lungs" in an unusual manner. The frog opens the nares of its nose, closes its glottis, and allows air into the buccal cavity. It then lifts the floor of the buccal cavity, thereby decreasing the volume of the chamber, closes its nares, and opens its glottis. Air is thus forced into its lungs. Exhalation is passive. Part of this process is similar to gill breathing.

But how did the frog's genome get rid of the operculum and gills of a fish, and form an appropriate mechanism for opening and closing the nares, and opening and closing the glottis, and shift the gill capillaries to lungs, and form functional appendages – all in a coordinated fashion? This is another case of irreducible complexity, because if one tried to produce an adult frog in a stepwise manner, the poor thing would not be able to ventilate its lungs or move about hunting insects on dry land.

If evolution/cumulative selection were to try to produce an adult frog first, rather than a tadpole, the problems would be even greater, because how would you now reintroduce a swimming tail and gills, get rid of the legs, and shift the lung's circulation back to the gills for the first tadpoles to emerge from the eggs laid by adult frogs – which would have contained the blueprints/coding for adult frogs – not tadpoles?

At this point, let's again reiterate that one gene equals one enzyme/protein; that one cell requires multiple types of proteins; that one tissue-type requires multiple types of cells; and that one organ requires multiple types of tissues. It is ridiculous to believe that so many "mutations" could have occurred on so many genes, as to produce the changes required for the transition of a fish to a tadpole, or to an adult frog. And, of course, the same would be true of the transition of a fish to a reptile, or amphibians to reptiles.

Now that we've discussed various types of ventilation, it is time to discuss the primary subjects at hand: namely, by what method could the lungs of amphibians, or reptiles, or their common ancestors, have transitioned to the unique type of lungs utilized by birds, or produced functional small caliber alveolar type lungs?

LET'S START WITH THE RESPIRATION in lungfish reptiles and amphibians, since birds are said to have arisen either from amphibians or a common-ancestor of both reptiles and birds – or possibly from dinosaurs.

The dinosaur problem needs to be settled and dismissed out of hand. In the first place, we have no idea how a dinosaur carried out respiration. All soft tissue evidence has long since disappeared. Moreover, Allan Feduccia, chairman of the department of biology at the University of North Carolina, Chapel Hill, and his colleagues, writing in the *Journal Science* in October of 1997, reported "unequivocal evidence" that birds could not have arisen from dinosaurs because birds lack the embryonic thumb of dinosaurs.

This concept was corroborated by the comments of J. A. Rubin of Oregon State University at Corvallis, when he stated the following in an Associated Press interview, also in 1997: "the idea that birds evolved from dinosaurs is definitely in trouble."

This would appear to put to rest the theory of John Ostrum (Yale University) that Archeaopteryx and Caudipteryx were descended from dinosaur lineages from which birds later branched. If nothing else, the biology textbook Biology Concepts and Connections, published in 2003, recorded the following: "Recently, researchers published a description of a 220 million-year-old fossil of a non-dinosaur reptile, Longisquama insignis, which appears to have feathers! Thus, the scientists argue feathers can no longer be considered a defining characteristic of dinosaurs related to birds. Of course, advocates of the bird-dinosaur link question this interpretation."

But what about those bird-dinosaur link advocates? Is there really hard evidence to support their opinions? Or, is it a matter of all such advocates having learned the same old evolutionary concepts by rote? Jonathan Wells answers the question in his book: *Icons of Evolution, Science or Myth?*

In his introduction, he points out the following: "When asked to list the evidence for Darwinian evolution, most people, including most biologists, give the same set of examples, because all of them learned biology from the same few textbooks."

The lungs of birds do not function on the basis of inhaling and exhaling tidal volumes of air. If this were the system of ventilation, as is the case in frogs, snakes, and human beings, up to one third of ventilation space in the trachea and bronchi could have been dead space, and thus an inefficient means of oxygenation – because exhaled air remaining in the dead space is inhaled back into the lungs during the next inspiration. This is one of the reasons that physicians do tracheostomies. The tracheostomies help to eliminate at least a part of the dead space.

Birds have a much more efficient method of ventilating their lungs. They use a continuous-flow system of fresh air moving in one direction. The system is ingenious and employs multiple laws of physics. The lungs themselves are anchored to the wall of the rib cage and move very little. Rather, air passes through small caliber tubules (air capillaries) running at right angles to pulmonary vascular capillaries. There is a continuous exchange of oxygen and carbon dioxide, without a decrease in the efficiency of gas exchange, as is the case in lungs with tidal exchanges. Birds do not have diaphragms. So, the thoraco-abdominal space functions as a single chamber.

As the sternum and fercula rock ventral-ward and forward, the ribs also move ventral-ward and forward increasing the volume of the thoraco-abdominal space. This produces a negative pressure, and outside ambient air is forced through the trachea, the primary bronchi leading to each lung, and through the intrapulmonary bronchus (also known as the mesobronchus) and partially into the dorsobronchi and air capillaries of the lungs themselves, and also into the abdominal and caudal-thoracic air sacs.

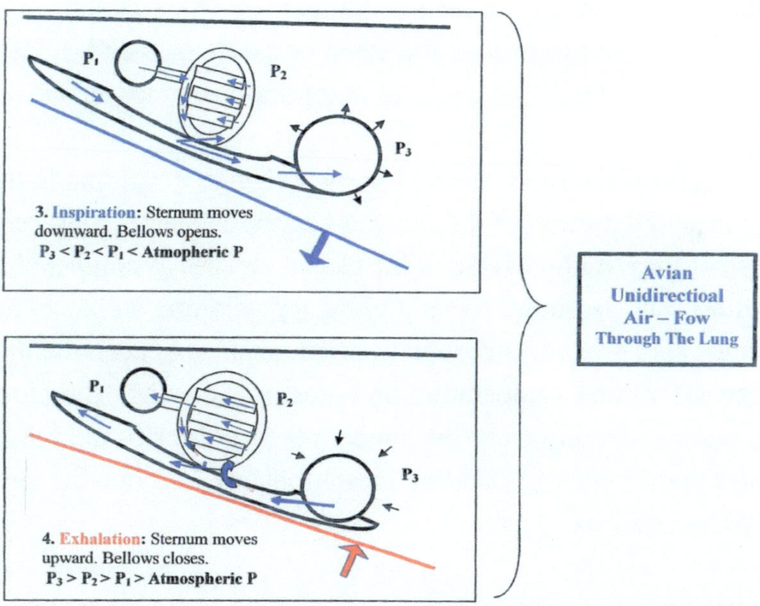

Diagram of Air Passing through a Bird's Lungs During Respiration

In addition to the paired caudal-thoracic and abdominal air sacs, there are also paired cranial-thoracic air sacs and a single intra-clavicular air sac. The caudal-thoracic and abdominal air sacs are like bellows, having drawn air in during inspiration, they push the same virgin air back through the mesobronchus, the dorsoparabronchi, and air capillaries of the lungs during expiration.

This forces the air that was already in the dorsobronchi and the air capillaries cranial-ward out of the lung and into the ventrobronchi in the cranial-thoracic and intra-clavicular air sacs, and also back into the cranial end of the mesobronchus and out the primary bronchi and trachea. Thus, with every inspiration and expiration, there is a constant unidirectional flow of fresh air through the air capillaries of the lung.

Air does not flow retrograde from the cranial-thoracic air sacs or the intra-clavicular air sac because the change in volume in the abdominal portion of the thoraco-abdominal cavity is greater than the change in volume in the cranial portion of the thoraco-abdominal cavity, where the cranial thoracic sacs and intra-clavicular air sac are located.

At constant temperature, a greater change in volume in the abdominal/caudal portion of the thoraco-abdominal cavity must produce a transient increased pressure in the caudal air sacs as compared to the cranial air sacs. Because, for any given temperature, a change in volume must also produce a change in pressure so as to keep the dynamics between volume temperature and pressure constant. Therefore, air flows unidirectionally from the caudal air sacs through the lungs and into the cranial air sacs and/or mesobronchus and out the primary bronchi and trachea.

NOW THE QUESTION ARISES how could any one mutation or any series of mutations on various genes of amphibians, who gulp air in tidal exchange volumes, or reptiles that also use tidal-volume-exchange alveolar type lungs, produce avian unidirectional constant flow type lungs in a stepwise and functional manner? <u>The answer is that it is absolutely impossible</u>!

There is nothing homologous to work with except for the trachea and the primary bronchi. Destroy or leave out one feature of the avian ventilation system, and you destroy the whole thing. It could not have been produced in a stepwise, chance manner. Any intermediate life form would have died trying to transform amphibian or reptilian ventilation systems into an avian type of ventilation, in a stepwise fashion.

This observation is another instance of irreducible complexity. If there are evolutionists who disagree, then let them lay out the pathway by which such an evolutionary transition could have been made, and kept the organism functional while transitioning from an amphibian or a reptile into a bird. The burden of proof lies with the evolutionists.

PEOPLE ARE USUALLY UNAWARE of their own breathing. That is because it's effortless, in the absence of disease. What happened to the problem with surface tension? Here is the answer: our bodies produce a group of amazing wetting-agents that not only markedly reduce surface tension, but actively prevent collapse of alveoli at the end of expiration.

Even without the problem of surface tension, if our alveoli were to collapse completely, significant energy would have to be expanded in trying to reinflate them with each subsequent breath. Every reader who has blown up a balloon has faced the problem that the body would face in trying to inflate a completely collapsed alveolus/balloon. The first efforts in trying to blow air into a collapsed balloon are met with severe resistance.

It is not until the balloon begins to inflate that the effort becomes easy. The difference is the change in surface area. The same pressure per square inch is now applied over a greater and greater surface area, and therefore the total force expanding the balloon is greater than when the balloon is collapsed. The same thing is true for the alveolus. The wetting-agents in human lungs not only minimize surface tension, but they resist compression and the complete collapse of an alveolus.

AS STATED ABOVE, the function of mammalian lungs requires the improbable production of a special group of phospholipid surfactant molecules: primarily dipalmitoyl phosphatidylcholine, along with five other phospholipids and four apoproteins. These special molecules, surfactants, can not only variably adjust surface tension, but, as stated above, they can, at the same time, resist compression. It is the presence of these molecules that allows us the luxury of relatively small lungs with tremendous surface area for exchange of carbon dioxide and oxygen, without the severe energy consumption problems that would normally be encountered in the presence of small diameter alveoli, air-liquid boundary layers, great surface tension, and therefore the need for great trans-pulmonary force in moving air in and out of the thorax.

This fluid lining is not only water. If it were, the alveolar wall tension would be much higher than it is, and large forces would be required to inflate the lung. This awkward situation is avoided by the presence of surfactants, lipo-protein complexes that bestow a very low surface tension on the liquid-air interface.

Surfactants are produced by Type II cells within the alveolar lining and have a half-life of about 12 hours in mammals. Surfactants are found in the lungs of amphibians, reptiles, birds, and mammals, and they may be present in some fish that build bubble nests. In humans, surfactant release is stimulated by sighing….

Alveoli fold as their volume decreases, and would become glued together by surface tension were it not for surfactant, which reduces the surface tension enough to allow easy inflation of the collapsed alveoli. When lung volume is reduced, extremely, the lung will collapse, a condition called atelectasis. Due to the presence of surfactants, an atelectatic lung can be re-inflated relatively easily.

Surfactants allow newborn babies to inflate their lungs. In mammals, surfactants appear in the fetal lung prior to birth. Newborns who produce no lung surfactants cannot inflate their lungs, a condition called neonatal respiratory distress syndrome.

Surfactants reduce resistance to blood flow by increasing the compliance of the capillary-alveolar sheet. When surfactants are present, capillaries are less likely to collapse, and once collapsed, they are easier to reopen.

Surfactants increase the osmotic pressure of the lung fluid, thereby reducing water flux across the lung epithelium.

The absence of the lipo-protein, phosphatidylcholine, is the primary cause of respiratory distress in premature infants. Without it, small alveoli collapse at the end of expiration, and have to be reinflated with each successive breath, thereby producing ventilation-perfusion mismatch right-to-left shunt of cardiac output, fatigue of the diaphragm and death.

A relative deficiency of surfactant, which leads to a decrease in lung compliance and functional residual capacity, with an increase in dead space, causes Respiratory Distress Syndrome (RDS). The resulting large ventilation-perfusion mismatch and right-to-left shunt may involve as much as 80% of cardiac output. Macroscopically, the lungs appear airless and ruddy (i.e., liver like). Thus, the lungs of these infants require a higher critical opening pressure to inflate. Diffuse atelectasis of distal air spaces along with distention of some distal airways and perilymphatic areas are observed microscopically. With progressive atelectasis, along with barotrauma or volutrauma, and oxygen toxicity, endothelial and epithelial cells lining these distal airways are damaged, resulting in an exudation of a fibrinous matrix derived from blood.

Thus it can be seen that we live only because this amazing system of molecules, surfactant is produced and secreted in the right place at the right time—stored and secreted by type II pneumocytes just prior to birth. This example alone is sufficient to fulfill the criterion set by Darwin to disprove his theory; namely, that alveolar lungs cannot be produced in a stepwise manner, the first alveolar type lung required the presence of surfactants of the same type found in current day alveolar type lungs.

Chapter Thirteen
The Human Knee

The manufacture of components for total knee replacement arthroplasties has taught us that the best design is that which simply resurfaces what "nature" provided as normal standard issue in the first place. Human engineering has not been able to improve on that design. The human knee, as designed, is already the ultimate in design considerations. Thus, with the above as an introduction, let's discuss the biomechanics of the knee in terms of prosthetic design considerations – as presented in Campbell's Operative Orthopedics – noting that the original design considerations for the human knee joint were the same in the beginning as they are now, and remain unchanged and unchanging:

"Anatomically, the knee is classified as a diarthrodial, more freely movable joint of the ginglynus or hinged type. However, several kinematic studies have confirmed that motion in the knee is not that of a simple hinge, but is an extremely complex series of movements about variable axes and in three separate planes during the course of a normal gait cycle. Flexion and extension do not occur about a fixed transverse axis of rotation, but rather about a constantly changing center of rotation, that is polycentric rotation. When plotted, the path of this changing center of rotation describes a J-shaped curve about the femoral condyles.

Flexion and extension of the knee are accomplished by both rolling motion and gliding motion between the femoral and tibial condyles. In addition to flexion and extension occurring in the sagittal plane, concomitant abduction and adduction are occurring in the coronal plane, and internal and external rotation are occurring in the transverse plane. Motion in all three planes has been measured during the normal gait cycle by several investigators using different techniques.

The findings in all the studies are basically the same and reveal approximate 70° of flexion and extension during the swing phase and 20° during the stance phase, approximate 10° of adduction and abduction, and 10 to 15 degree of internal and external rotation during each gait cycle. Kettlecamp et al. performed kinematic studies of the knee during selected activities of daily living and found that the knee motion required was activities specific and in certain activities could be significantly greater than the above values for walking over level surfaces at a normal cadence.

Kinematics while important is only one aspect of the biomechanics of the knee that relate directly to the success of arthroplasty. Knowledge of the magnitude and distribution of forces across a normal knee joint in a variety of positions and activities is essential in prosthetic design and development. Morrison has calculated the loads applied to the knee joint during level and inclined walking and while ascending and descending stairs. He found that in level walking the joint surfaces are subjected to a loading force equal to three times the body weight. When a person is negotiating ramps or stairs, the force could increase to over four times the body weight." (Campbell's Operative Orthopedics, 1992, page 391)

"Although the ideal prosthesis has not yet been designed, certain criteria for this elusive ideal are now being mapped. It appears desirable to design components that allow a normal or nearly normal range of motion in all three planes of knee motion, that is, flexion and extension, adduction and abduction, and rotation. They should provide a constantly changing center of rotation, as does the normal knee.

These criteria are best met by constructing femoral runners that are convex in two planes and that articulate with tibial surfaces that are concave in two planes, the total design being made to conform to the normal condylar curves in the knee, including the menisci. The design should allow preservation of the normal ligaments whenever possible to provide for balancing the soft tissue tension by appropriate surgical releases." (Campbell's Operative Orthopedics, page 393)

AT THIS POINT, HAVING discussed the structure of bone and the kinematics of the knee joint, it is time to begin our discussion of the elements of design inherent in the form and function of the knee joint, by asking the reader to consider how an engineer might go about solving the following functional problems involved in biped weight-bearing and motion, or how the reader might go about solving the following problems, or how those who believe in evolution might go about solving the problems involved with biped locomotion. Then, after consideration, consider again, and see if anyone can really believe that the human knee joint is the product of random-chance evolution. No, intelligent design answers all the problems beautifully, using elegant engineering principles. Here are some of the challenges that design had to solve:

- Develop adequate torque, while minimizing the power required to straighten the knee joint against resistance, in varying degrees of flexion.
- Minimize the need to shift the center of gravity of the body to the right or left in order to maintain balance during biped ambulation.
- Minimize and counteract the lateral forces produced in response to shifting the knees towards the body's midline during

weight-bearing – so required in order to decrease the work-requirement in standing and walking erect.

- Accommodate to uneven terrain, requiring the joint not only to flex and extend, but also to rotate inwardly and outwardly, and deviate into varus and valgus configurations – in order to absorb the various stresses and strains placed upon the joint – while traversing uneven terrain.
- Produce a joint that is rigid in extension, so that there is stability as one takes a step, but is relaxed during flexion, so as to minimize wear and friction and reduce the power required to produce rapid flexion.
- Convert compression forces into tensile forces so as to limit deformation and injury to the elastic cartalenagous surfaces of the femur and tibia - and at the same time distribute incident forces over as wide a surface area as possible – in order to reduce the load per unit area on the femoral and tibial surfaces and, therefore, reduce the likelihood of wear and degenerative change.
- Diminish the increased load on the lateral femoral condyle and the lateral tibial plateau of the lateral joint compartment of the knee – subsequent to the necessity of deviating the knees towards the midline of the body, as a necessity in minimizing the energy requirements of biped walking.
- Prevent jamming of the joint surfaces during flexion and extension of the knees and maximize lubrication.

All of the above design considerations were solved beautifully, with each element of the inherent design considerations integrated into a perfectly functioning unit. How can the interconnected solutions to the above engineering concepts be explained on any basis other than simultaneous occurrence and designed irreducible complexity? The dogma of evolution can in no way explain how the various design components could have been selected on a stepwise basis.

The reason is this: individual generations, or individuals within a generation, could not have tested out the various optional engineering concepts – and then had them selected. Because evolutionary hypothesis has long since discredited Lamarckian inheritance of acquired characteristics, and stepwise construction of a biped knee joint could only have resulted in early degenerative change, and/or dislocation, and/or significant dysfunction, and, therefore, adverse selection – assuming, for the moment, that natural selection and evolution might be true.

Upon which genes, and upon which chromosomes, would random chance evolution have carried out point mutations, or insertions, or deletions, or chromosomal recombinations, in order to bring about the various engineering changes required to transform a straight-legged quadruped knee into a valgus biped knee? We must ask evolutionists to show us where the increased chromatin came from to accommodate the increased complexity in a biped knee joint, as compared to a quadruped joint; not speculate on a possible explanation.

IN ORDER TO INCREASE TORQUE, the distal end of the femur terminates in a pulley-like, expanded, rounded flair. This moves the effective lever arm of the forces acting on the tibia (the force generated by the quadriceps muscle) by elevating the quadriceps mechanism away from the center of motion of the knee – thus lengthening the lever arm. During flexion and extension, the patella can increase the mechanical advantage of the knee by 50 percent. Thus, the size of the quadraceps muscle, and its energy requirements, can also be decreased by 35 to 50 percent. Steindler (p. 23)

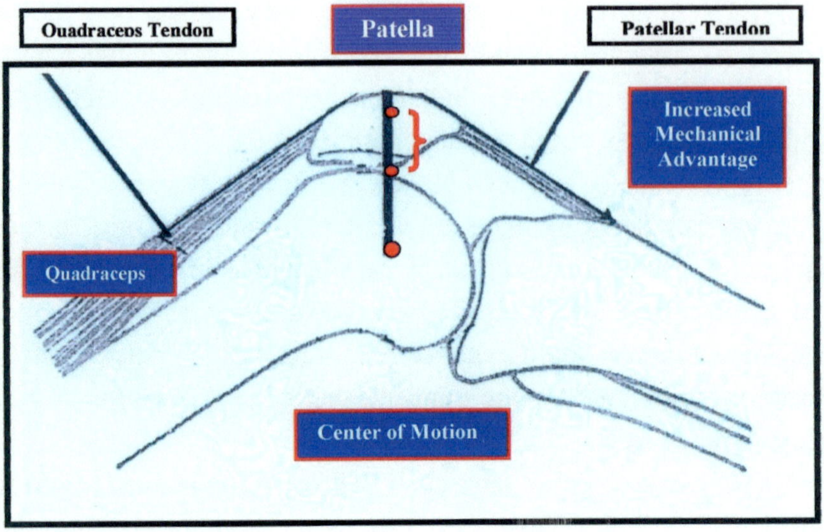

The biomechanical principles employed above produce an increase in the knee joint torque by approximately 50%, without increasing the size of the quadriceps muscle. Nice engineering!

In addition, adding a patella into the quadriceps tendon further elevates the tendon away from the center of motion of the knee, and further increases the joint reaction force by an additional 50 percent.

At the same time, that torque is increased, friction is decreased, by placing the hyalin cartilage surface of the patella against the hyalin cartilage pulley-like surface of the femur, separated by a thin film of synovial fluid, thus reducing the coefficient of friction, in the presence of synovial fluid, to 10 times less than ice on ice. This avoids the problem of having a quadriceps tendon or patellar tendon running in the trochlear groove of the femur. Tendon is good for activating joints because of its excellent resistance to tensile forces. But tendon is not a good material to pull around corners, because of its wear characteristics, and because of its coefficient of friction. Tendon is very sensitive to friction and to compression (Ficat and Hungerford, page 23).

Without the patella, the patellar tendon would be compressed against the femoral condyles during both flexion and extension – markedly increasing the possibility of quadriceps mechanism rupture, as well as wear in the trochlear groove. Again, beautiful engineering! Can anyone truly believe that random chance evolution could integrate all of these engineering principles into the function of the patellofemoral portion of the knee joint?

WHEN A BIPED ANIMAL, such as a human being, stands on one lower extremity, it is necessary to bring the center of gravity of the body over the support point. If the center of gravity is not maintained over the support point, the individual involved will fall either to one side or the other, or forward or backward. Walking motion is accomplished by exercising a continual controlled fall, in the direction of motion. When ambulating forward, the pelvis and body are allowed the fall a short distance forward, toward the ground.

The lower extremity opposite the one on which the body is balanced during the fall is brought forward to stop the fall, and, in so doing, becomes the new balance point. Thus, the pelvis of a biped in motion describes a sort of shallow, up-and-down, sine wave, as an individual proceeds forward.

On the other hand, the pelvis remains relatively level with regard to side to side motion. Intuitively, if either the left or right hips of a biped were the true balance points, the center of gravity would have to swing back and forth, first over one hip and then over the other hip.

This type of motion is called a waddling gait. It is the way people are forced to walk when they have weak abductor muscles, or abnormalities of their hips or knees. The energy expended when using a waddling type gait is far greater than the energy requirements for normal human gait.

Fortunately for human beings, intelligent design was aware of the problem that would be encountered if a biped had to walk on the type of lower extremities utilized by quadrupeds, namely a lower extremity that simply extends straight downwards from the hip joint.

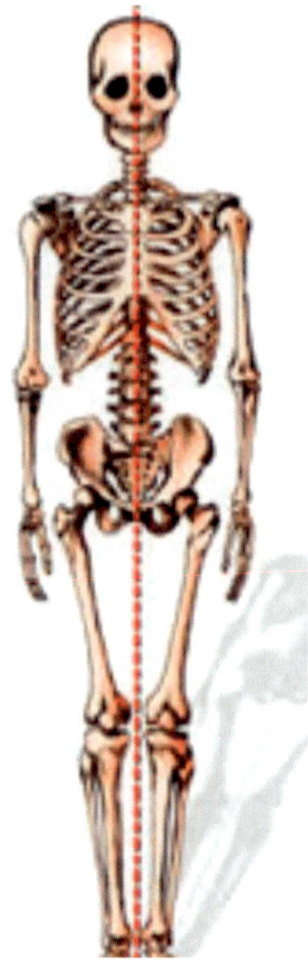

Notice how the legs of humans move toward the center at the knees. This design prevents the human from waddling as the person walks or runs. Interestingly, no other species has this marvelous ability to walk without waddling. Also notice that in moving close to the mid-line that the weight of the body balances on one leg in the mid-line as a person walks or runs. People who don't have this accommodation waddle in walking or running.

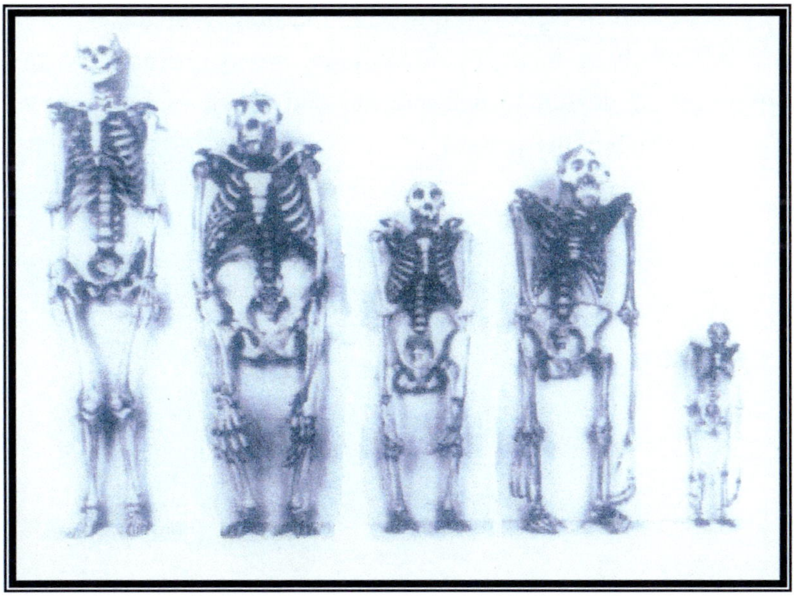

Knee-Valgus man Gorilla Chimpanzee Orangutan Gibbon

Note the valgus angle in the knee of the human only.

If human beings had quadruped type lower extremities, they would be forced to walk with a waddling type gait. Intelligent design avoided the necessity of using a waddling gait by deviating the hip joint and femur toward the midline, and then angling the tibia straight downwards, in order to plant the distal weight-bearing surface of the ankle parallel to the ground.

In this way, the actual balance point when standing on either the right or left lower extremity becomes either the right or left knee, shifted toward the midline, therefore allowing the center of gravity of the body to be balanced with ease and with little expenditure of energy in maintaining the center of gravity over the knees, acting as balance points – rather than the hips.

The reader can check this out. Just stand in front of a mirror and shift your weight back and forth from one foot to the other. This is accomplished with ease. Note how slight a shift of the center of gravity to the left or right is required to maintain balance.

All of these intricate movements are coordinated with great complexities of design to achieve the end result of body movement with ease in the healthy person.

Chapter Fourteen
Summary of Intelligent Design

As was stated in the Preface for those who claim to be scholars, to ignore fact is foolish – especially scientific fact. It is strange that those who indulge in the theory of macroevolution ignore the facts of science. They seem to enjoy some form of self-delusion.

Throughout this book, part of the intent of the writings has been to show scientific criticisms of evolution. However, it is important to show how scientific facts demonstrate that the earth and the various species on it have been designed for special needs and purposes.

It is interesting to note that there has been a shift by some educators to balance their curriculums by not only covering the concepts of macroevolution, but to teach the concepts of intelligent design.

At Lakeview School, Rodney LeVake was reassigned after he allegedly failed to adequately cover the curriculum requirements for evolution. He said that he intended to teach scientific criticisms of evolution and that he would "accompany [the] treatment of evolution with an honest look at the difficulties and inconsistencies of the theory without turning my class into a religious one."

There is a strong movement away from macroevolution toward other theories of how earth's species came about. In the 1950s, the majority embraced the macroevolution concepts, but today this is changed, now the majority of scientists have moved away from macroevolution to other concepts, including intelligent design. They have done this in part due to flaws and speculation by those who preach macroevolution as fact, when in reality it is not.

The following quote summarizes the concepts of Darwinian evolution. It was included in the writings of Granville Sewell, a professor of mathematics at the University of Texas: "In this wide-ranging collection of essays, Sewell concludes that while there is much in the history of life that seems to suggest natural causes, there is little evidence to support Charles Darwin's idea that natural selection of random variations can explain major evolutionary advances."

We have only touched on the complexities of the human body, which include much of our distress, but hopefully enough has been presented to make the case that the species of earth are far too advanced for each tiny item in its place to have caused the whole animal kingdom of earth to function properly. Anyone who has lost any use of a part of the body understands the beauties of a healthy body that has been skillfully designed to allow a person to function at normal tasks. If the world in which we live were truly formed by evolution, wouldn't one expect evolution to still be in progress? And wouldn't it be expected that we would still see examples of only partly completed animals and plants, only partly adapted to the world in which we live?

SOME SCIENTISTS, SUCH as Professor Hubert Yockey, University of California, arbitrarily choose to explain their evolutionary hypotheses in terms of mathematical aberrations within the limits of information theory. For those who choose to consider evolution a process of random modification of genetic information via the tenets of information theory at the molecular level, mathematician William Dembski summarizes the problem with that approach as follows: "Let's examine whether chance is capable of generating complex specified information.

First, notice that pure chance, entirely unsupplemented, and left to its own devices, is incapable of generating complex specified information. Chance can generate complex unspecified information, and chance can generate non-complex specified information. What chance cannot generate is information that is both complex and specified." Yet that is exactly what is required by the concept of evolution.

Shannon information theory is counterintuitive. The longer the message, the more information it contains, but the greater the chance for error, and therefore the less specific a message becomes. In biology, the information passing from generation to generation is staggering.

Each chromosome is millions of base-pairs long. Thus, an error-rate of only one in 10,000 turns into a significant number of errors. But, biological information is not transmitted. It is inculcated. Therefore, information theory does not apply. Each new molecule of DNA or RNA is manufactured by building it on a pre-existing molecule acting as a template. Each cell has built in "repair processes" which further insure that information is inculcated properly.

Information theory applies only when information transmission systems (signals) employ: a source, a transducer, a transmitter, signal noise, a receiver, a receiver-transducer, and a need to decode and interpret the transmitted signal. In biology, there is an end product, not a message transmission.

The end product is an inculcated duplicate gene. When discussing gene reproduction, and hence macroevolution, we are not talking about the production of various types of RNA or protein. We are discussing only the duplication of DNA. There is no signal to interpret. There is only duplication. True, there may be minor "typographical/stamping errors," but if the new, duplicated gene allows the production of a viable phenotype, the "stamped-out," inculcated "message" contained within the new DNA molecule remains within the limits of quality-controlled duplication. Remember, 99.9% of copying errors are detrimental and eliminate themselves.

Therefore, the duplication process that occurs within the replication-forks of DNA, duplicated gametes – that carry the germ-line – produces essentially exact copies – just as one makes a copy of a key;

not as one sends a radio message, or a telegram, or transmits a television program. Again, remember that when we talk about macroevolution, we are talking about the reproduction of genes and the reproduction of DNA (inside the catalytic space of DNA polymerase). When we discuss the concept of evolution, we are not talking about the processes that occur within the catalytic space of the DNA transcription bubbles responsible for the production of RNA or structural or enzymatic proteins.

Dembski is right - the inculcated information, transferred by the duplication process within the carefully controlled catalytic space of DNA replication forks, is not only complex, but absolutely specific as to the process that results in correct protein reproduction and correct protein folding.

Thus again, the paraphrasing of Hugo DeVries best summarizes the problem with recursions, fractals, and stochastic processes as a method which might allow chance to produce the complexity and specificity that constitute human beings. It is not the process of duplication which must be explained by evolutionists, it is the origin of the process itself that must be explained, on the basis of chance. Design and inculcation explain both origins, easily.

The questions we asked can be condensed and rephrased as follows: given an adequate supply of inorganic substrate, which process is more likely to produce a specific functional biologic system - chance or design? Notice that we did not ask what methods or laws chance might employ in reaching a stated goal, nor did we ask what methods design might employ in reaching the same goal. So, the methods used by chance, whether it be mutation in combination with trial and error, or recursion, or a stochastic process (which is really only a combination of trial and error and recursion), becomes irrelevant.

It makes no difference what methods chance or design might employ in reaching a goal, nor does the time required, nor the energy expended. The only thing that matters is whether or not the answer to our question lies within the constraints of probability.

Again, as stated previously, probability doesn't care about the process used, the time spent, or the energy expended. Probability cares only about the end result and the number of elements to be considered in the final result.

So, at this point, we can state that we have now considered recursion, fractals, and stochastic processes in general. However, the additional considerations have only confirmed the answer that the odds ratio, alone, had already provided. The odds ratio tells us very clearly, from the start, that the likelihood of the biologic systems we discussed are many times more likely to have been produced by design than by chance. The ratio is:

$$\frac{1}{8.3 \text{ X } 10^{62}}$$

Thus the basal ganglia system is 120 million, billion, billion, billion, billion, billion, billion times more likely to have occurred by design than by chance.

There are limits to error. It was not necessary to employ the statistical methods used to determine reliability, because we did not ask the question: to what degree of certainty are our conclusions correct? Rather, we asked only the simple question: what is the likelihood of the biologic system we discussed having been produced by chance, as compared to design? The only relevant factors that determined our answer were the laws of probability and the odds ratio.

We have pointed out the lungs are physiologically and mathematically irreducibly complex.

We have also pointed out that the neuromotor system of the human body, and the genes that produced it, are not only complex, but functional, and irreducibly complex. Should one remove any one component of the neuromotor system, it will not function with smooth and fluid motions. Instead, diseases such as Huntington's chorea, Parkinson's disease, ballismus, tremor, past-pointing, limping, spasticity, etc. are seen to develop.

These types of abnormal function would absolutely interfere with fitness and reproductive advantage. The conclusion of our analysis of chance stochastic processes resulting in the production of lungs or the human neuromuscular system fulfills Darwin's criterion for the evidence that he stated would prove his postulate wrong! Professor Behe has. Professor Dembski has. And, we have.

Selection has to take what it gets. And what it gets is normally worse than what it had. It is hard to climb an evolutionary ladder with that kind of scenario, against the limitations of probability. DeVries may have been wrong about mutations, but may we state again that he was absolutely right about the shortcomings of Darwinism when he stated that natural selection might explain survival of the fittest, but it couldn't explain arrival of the fittest.

A particularly unfortunate distortion of fact is the manner in which evolutionists misstate and misapply the Hardy-Weinberg Law. It is disturbingly inappropriate that evolutionary biologists have produced a version of the law modified to suit the needs of macroevolution.

Now, we are free to give those who believe in evolution both infinite time, infinite numbers of trials, and infinite methods of production of a given result – and still – science demands that in the case of the swallowing mechanism, the probability of the mechanism having developed by chance is zero, and, how much the less for the human body as a whole!!!

Few readers would be willing to bet their fortunes on a lottery ticket where the odds against winning are only a few billion to one. So, why would anyone want to bet a correct understanding of life on chance evolution – where the odds of being correct are far, far less; actually zero, from the standpoint of pure science.

IT MUST BE CONCLUDED THAT DARWIN'S OWN CONDITION of trial and error, and natural selection, failing to produce any biologic system, has indeed falsified his hypothesis of macroevolution. Microevolution is a fact. Macroevolution is a highly unlikely hypothesis. Darwinian macroevolution should be presented only in conjunction with its shortcomings and contradictions.

"Intelligent design is a framework of logical thinking based on the observable axiom that can be used to analyze scientific data. This book promotes intelligent design because it is in many ways superior to the theory of evolution." (Article in *Discover* from the Web site)

There is a clear-cut irrefutable answer here. The overwhelming probability is that the systems discussed, and therefore, by extension, human beings, are the result of design, not chance. And, that the negative probability associated with a stochastic origin makes our answer even more certain.

Read what Albert Einstein's biographer said:

"... although Einstein did not believe in God, he seems to [have] fruitfully treated the world as if it were created by a deity:" Note what his biographer Banish Hoffman wrote the world famous scientist: 'When judging a scientific theory, his [Einstein's] own or another's, he asked himself whether he would have made the universe in this way had he been God. This criterion . . . reveals Einstein's faith in an ultimate simplicity and beauty in the universe. Only a man with a profound religious and artistic conviction that beauty was there, waiting to be discovered, could have constructed theories whose most striking attribute, quite overtopping their spectacular successes, was their beauty.'" (Banish Hoffman, Albert Einstein: Creator and Rebel, 1972)

In ending this small acknowledgement of man's place in the universe, the author would like to state the following:

Man and woman were designed to grow from childhood to adulthood, learning about the earth, the universe, and the art of living with the largest brain of any animal, opposable thumbs to handle things, and valgus knees set close to the mid-line so as to walk uprightly without a waddling gait. There is also an amazing match of integrated male and female bodies, so that the chemistry and function of their bodies allows the conception of children with the sealing off of one male sperm and one female ovum instantaneously, so that each human being can be developed with a specific set of characteristics we think of as an individual specific human being.

END

Man and Woman – An Amazing Match

Additional Notes

A Critical Evaluation of Intelligent Design:

- Change over time, history of nature, any sequence of events in nature
- Changes in the frequencies of alleles in the gene pool of a population
- Limited common descent: the idea that particular groups of organisms have descended from a common ancestor
- The mechanisms responsible for the change required to produce limited descent with modification, chiefly natural selection acting on random variations or mutations
- Universal common descent: the idea that all organisms have descended from a single common ancestor
- Blind watchmaker thesis: the idea that all organisms have descended from common ancestors through unguided, unintelligent, purposeless, material processes such as natural selection acting on random variations or mutations
- Darwinian mechanism of natural selection acting on random variation, and other similarly naturalistic mechanisms, completely suffice to explain the origin of novel biological forms and the appearance of design in complex organisms. This is often referred to as the Theory of Evolution
- Microevolution - changes in the genetic make-up of a population
- Macroevolution - perceived linear change between the higher taxonomic levels
- Speciation - change from one species into another species
- *From Robin Collins*

Are Chimps and Humans Really All that Much Alike?

A popular Darwinian meme is that human and chimp genomes are ninety-something percent identical. It varies a bit, but usually hovers close to 99%. The meme hides all sorts of assumptions, of course, but the take home lesson for the headline reader is plain enough: we're almost exactly the same as chimps.

Though the 99% number has received some qualifiers, and has even been referred to as a "myth" in *Science*, the basic idea remains firmly entrenched in the media collective consciousness.

But evidence seems to be piling up that the similarities are not nearly what has been advertised. Geneticist Richard Buggs has reflected on this, and has even predicted "that when we have a reliable, complete chimpanzee genome, the overall similarity of the human genome will prove to be close to 70% (and very far from 99%)."

It will be interesting to see how Buggs' prediction holds up over time. If he's right, this will be one more switch from "meme" to "myth" in the Darwinian ledger.

I should confess that I haven't followed this debate closely, simply because I don't think that the meme, even if true, really shows much. Here's what I mean. Some years ago, a reporter called the Discovery Institute asking for a comment on a new study showing a 75% genetic similarity between human beings and a nematode (if I remember correctly). The reporter asked me what I thought about the discovery that we were 75% identical to a nematode. I suggested that there was a difference between our genomes being quite similar to the nematode's, and human beings being quite similar to nematodes. That didn't seem to connect, so I said: "Well, why doesn't the story cause you to reassess the assumption that we're basically identical with our DNA? If somebody told me we were 100% chemically identical with chimps, I would conclude that we must be a lot more than mere chemicals, since chimps and humans are quite different. Now since humans and nematodes are obviously quite different, while our genomes are similar, doesn't that suggest that genetics tells us less about ourselves than we had assumed?" I'm pretty sure my comments didn't make the story.

Jay Ricards, 2010

Intelligent Design Studied

The theory of intelligent design (ID) holds that certain features of the universe and of living things are best explained by an intelligent cause rather than an undirected process such as natural selection. ID is thus a scientific disagreement with the core claim of evolutionary theory that the apparent design of living systems is an illusion.

In a broader sense, intelligent design is simply the science of design detection -- how to recognize patterns arranged by an intelligent cause for a purpose. Design detection is used in a number of scientific fields, including anthropology, forensic sciences that seek to explain the cause of events such as a death or fire, cryptanalysis and the search for extraterrestrial intelligence (SETI). An inference that certain biological information may be the product of an intelligent cause can be tested or evaluated in the same manner as scientists daily test for design in other sciences.

ID is controversial because of the *implications* of its evidence, rather than the significant *weight* of its evidence. ID proponents believe science should be conducted objectively, without regard to the implications of its findings. This is particularly necessary in origins science because of its historical (and thus very subjective) nature, and because it is a science that unavoidably impacts religion.

Positive evidence of design in living systems consists of the semantic, meaningful or functional nature of biological information, the lack of any known law that can explain the sequence of symbols that carry the "messages," and statistical and experimental evidence that tends to rule out chance as a plausible explanation. Other evidence challenges the adequacy of natural or material causes to explain both the origin and diversity of life.

Intelligent design is an intellectual movement that includes a scientific research program for investigating intelligent causes and that challenges naturalistic explanations of origins which currently drive science education.

Intelligent Design Network

As an additional closing comment, it must be remembered that chimpanzees do not have valgus oriented knees. Thus they, like all other animals of earth, except humans have lower extremities that simply come straight down from the hip. Accordingly they must walk on their lower extremities in a waddling type gait. This is a feature totally avoided in human beings, the only animal on earth that walks uprightly.

Author

References

1. Adams, M.: The Genome Sequence of Drosophila melanogaster; Science
2. Anderson MD, J. E.: Grant's Atlas of Anatomy 7th Ed.; Williams and Wilkins Co. Baltimore; 1978
3. Andersson, D. I, and Huges, D.: Muller's Ratchet Decreases Fitness of a DNA Based Microbe; Proceedings of the National Academy of Sciences of the USA; volume 93, January 23, 1996
4. Ayala, F.: in Creative Evolution; Campell, J. and Schopf, J. 1994; Jones and Bartlett
5. Bechtol, C. O.,Ferguson, A. B., Laing, P. G.:Metals an Engineering in Bone and Joint Surgery; Williamson Wilkins, Baltimore, 1959
6. Behe, M. J.: Darwin's Black Box; Simon and Schuster, New York; 1996; p 16
7. Bennett, D. J.: Randomness; Harvard University Press, Cambridge Mass., Paperback Edition, 1999
8. Berkow MD, R. and Fletcher MB, A. J.: The Merk Manual 16th Ed., Merk Research Labaratories. Rahway N. J. 1992
9. Berne, R. M., and Levy, M. N.: Physiology, 4th Edition, Mosby, St. Louis, Philadelphia, London, 1998
10. Borel, Emile: Probabilities and Life; M. Baudin, New York; 1962
11. Bosma, J. F. Deglutition Pharyngeal cap stage; Physiological Preview No. 371957
12. Bunnell, S. Surgery of the Hand, 3rd Edition: Lippincott, Philadelphia
13. Burle: Electro-Optics Handbook, 1989; Burle Technologies
14. Campbell, N.A., Rees, J. B. Mitchell, G.M. and Taylor, M.R.: Biology, Concepts and Connections 4th edition; Benjamin Cummings, San Francisco, 2003
15. Carr, Joseph: A Crash Course in Statistics Hightext Solana Beach CA; 1994
16. Colvill, T., and Bassert, J. M.: Clinical Anatomy and Physiology for Veterinary Technicians; Moseby, St. Luis, 2002

17. Cox, J. S., et al: The Degenerative Effects of Partial and Total Resection of the Medial Meniscus in Dog's Knees; Clinical Orthopedics, 1975, vol. 109
18. Crenshaw MD, A. H.: Ed. Campbell's Operative Orthopedics; 8th Edition; Mosby, St. Louis, Baltimore, Boston, Chicago, London, Philadelphia, Sydney, Toronto
19. Crozet, Y Referenced by National Solar Observatory, Sacramento Peak NM;
20. Dawkins, R.: The Blind Watchmaker; W. W. Norton & Co., N. Y., 1996
21. Dawkins, R.: Climbing Mount Improbable; W. W. Norton & Co., N. Y., 1997Fronhoffer, H. and Nusslein – Vollard, C.: Organization of Anterior Patterns in the Drosophila embryo by the maternal gene Bicoid; nature, 1986, 324
22. Dembski, W.A. and Ruse, M. Editors: Debating Design; Cambridge University Press; Cambridge, 2004
23. Dembski, William A.: Intelligent Design, 1996; Inter Varsity Press, Downers Grove, Illinois
24. Doty, R.: Neural Organization, section 6, Elementary Canal; volume 4; American Physiological Society, Washington
25. Duarte,E.A., Vovella,I.S., Ledsma,S, et al: Subclonal Component of Consensus Fitness in an RNA. Virus Clone; J. Virology; 68 (7), July 1994
26. Dyson, F.: Origins of Life; Cambridge University Press, 2nd edition, 2000; Cambridge, United Kingdom
27. Estridge N.; Gould S. J., 1972: Punctuated Equilibria: an Alternative to Phyletic Gradualism. In Schopf, T. M. (ED.) 1972: Models in Paleobiology. Freeman Cooper
28. Ficat, R. P. and Hungerford, D. S.: Disorders of the Patellofemoral Joint, Williams and Wilkins Company, Baltimore, 1977
29. Gamlin, L. and vines, G: editors The Evolution of Life; Oxford University Press, New York; 1991, p.6
30. Gerhard, J. and Kirschner, M.: Cells, Embryos, and Evolution; Blackwell, 1997; London
31. Gillespie, J.H Molecular Evolution and the Neutral Allele Theory: in Oxford Surveys in Evolutionary Biology; Harvey P. H., Partridge,L., 4; 1987

32. Goldschmidt, R.: Material Basis of Evolution; (Silliman Milestones in Science); Yale University Press, New Haven CT 1982

33. Gould, J.L. and Keeton, W.T.: Biological Science, 6th edition; W. W. Norton & Co., New York; 1996

34. Guyton's, Basic Human Physiology Saunders 1971

35. Guyton's Textbook of Medical Physiology; Sanders Company 1976

36. Helfet, A. J.: Disorders of the Knee; J. B. Lippincott Co., Philadelphia, 1974

37. Heppenstall, R. B.: Fracture Treatment And Healing, WB Saunders Co., Philadelphia, 1980

38. Hiemae (K. M.: Mobility and Mastication – a review of the activity of the jaw muscles and the movements they produce in chewing; Development, Function, and Evolution of the Teeth, 1978; Academic Press, New York and London

39. Hoel, P;G;, Port, S.C and Stone,C.J.: Introduction to Probability Theory; Houghton Mifflin Co., Boston; 1971

40. Jaisingh, L. R. ; Statistics; McCraw-Hill, New York San Francisco Washington D.C.; 2000

41. Kauffman, S.A: Investigations; the Nature of Autonomous Agents and The Worlds They Mutually Create; Santa Fe Institute Preprint

42. Kenyon, D. H., and Thaxton, C. B: Of Pandas and People, 4th edition; Haughton Publishing Company, Dallas TX, 1999

43. Kimura, M: The Neutral Theory of Molecular Evolution: A Review of Recent Evidence; Japanese Journal of Genetics 66, 1991

44. Krause, W. R., Pope, M.H., Johnson, R. J., and Wilder, D. G.: Mechanical Changes in the Knee After Minisectomy; J. Bone and Joint Surgery Am. 1976, Vol. 58

45. Lewin, Benjamin: Genes VII; Oxford University Press, Oxford, 2002

46. Lewin, R: Patterns in Evolution, The New Molecular View; Scientific American Library, N.Y. 1999

47. Lewontin, R.C.: cited in: Electrophoresis in the Development of Evolutionary Genetics; Genetics, 128; 1991

48. Maly, I.V.: Diffusion Approximation of the Stochastic Process of Microtubules Assembly; Bulletin of Mathematical Biology; No. 64; 2002

49. Margulis, L, and Sagan, D: Acquiring Genomes, A Theory of the Origins of Species; Basic Books, first edition, 2002

50. McNicho, D. and. Roughly, P. J.: Extraction and Characterization of Proteoglycans from Human Meniscus; J. Biochemistry Vol. 185 1980

51. Meyer, S. C. The Origin of Biological Information and the Higher Taxonomic Categories; Proceedings of the Biological Society of Washington, 117 (2): 2004

52. Miller, R.F.1994. The physiology and morphology of the vertebrate retina. In S. J. Ryan, editor Retina, second edition, volume 1: Basic Science in Inherited Retinal Disease (St. Louis: Moseby)

53. Nelson, DL, and Cox M.M: Lehninger's <u>Principles of Biochemistry</u>; Worth Publishers, 2003

54. Pennock, R.T. Editor: Design Creationism and its Critics; MIT press, Cambridge; 2001

55. Perakh, M.: Unintelligent Design; Prometheus books 59 John Glenn Dr., Amherst, NY 2004

56. Pramanik, A.: Respiratory Distress Syndrome; e-medicine Editors: Donn, S.M.; Konop, R.; Clark, D.A.; and Wagner, C.L.: 7-2-02

57. Randall, D., Burggren, W., and French, K.:Ekert's Animal Physiology 5th edition; W. H. Freeman & Co., New York, 2002

58. Rhoads, R., Pflanzer, R.: Human Physiology, 4th edition: Thompson; Brooks/Cole, Pacific Grove California; 2003

59. Rielly, D.T.; and Martens, M.: Experimental Analysis of the Quadraceps Muscle Force and the Patello- femoral Joint Reaction Force for Various Activities; Acta Orthopedica Scandinavia 43, 1972

60. Schrodinger, Erwin: What is Life? Cambridge University Press, 2000: Cambridge Massachusetts

61. Schutzenberger, M-P.: The Return of the Hopeful Monsters; Natural History Magazine: 86, June 1976 Schutzenberger, M-P: The Miracles of Darwinism and Interview with Marcel-Paul

Schutzenberger; *La Recherche*: (French) La Recherche, N: 283: Jan. 1996

62. Seedhom, B. B., Dowson, D. and Wright, V.: Functions of the Menisci. A Preliminary Study,; J. Bone and Joint Surgery. Br., 1974, Vol. 56
63. Serway, R. A. and Jewett, J. W. JR.; Principles of Physics, 3rd Ed. Brooks/Cole- Thompson Learning, 2002
64. Shannon C. E: A Mathematical Theory of Communication, The Bell System Technical Journal; Volume 27, July-October 1948
65. Shrive N.: The Weight-bearing Role of the Menisci of the Knee: Proceedings British Orthopedic Research Society; J. Bone and Joint Surgery, 1974, Vol. 56B
66. Simmons, G.: What Darwin Didn't Know, A Doctor. Dissects the Theory of Evolution; Harvest House Publishers, Eugene Oregon; 2004
67. Smith, M.: The Theory of Evolution, Canto Edition; Oxford University Press, Cambridge; 2000
68. Stearns, S.C. and Hoekstra, R.: Evolution an Introduction; Oxford University Press Inc., New York; 2002
69. Steindler, A.: Kinesiology of the Human Body; Charles C., Thomas, Springfield, 1955
70. Stewart, James: Calculus 5th Edition, McMaster University, Thompson Brooks/coal, United Kingdom
71. Tigner, P.M.: Personal Communication, Scripps Institute of Oceanography, La Jolla California; October 24, 1989
72. Turek, S. L.: Orthopedics Principles and Their Application, third edition, J. B. Lippincott Co., Philadelphia, 1977
73. Watson, J., Baker, T., Bell S., Gann, A., Levine, M., Losick, R.: Molecular Biology of the Gene, 5th Edition, 2004; Pearson Education Inc., Benjamin Cummings, CSHL Press San Francisco California
74. Weissleder, H. and Schuchhardt, C.: Lymphedema; Viavital Verlag, Wesel, Baden Baden, 2001
75. Wells, J.: Icons of Evolution– Science or Myth, 2000; Regnery Publishers/Eagle Publishers, Washington DC
76. Williams, G.C. 1992 Natural Selection: Domains, Levels, and Challenges. Oxford: Oxford University press

77. Williams, P. L. and Warwick, R.: editors: Gray's Anatomy, 36 edition; W. B. Saunders Company, Philadelphia; 1980

Glossary

Actual Design
A phenomenon which is "universally" recognized as truly designed; verified teleology.

Apparent Design
Objects which contain features that most observers would acknowledge as giving the appearence of design.

Artifactual Design
Any phenomenon or event embodying counterflow.

Artificiality
A phenomenon that exhibits intelligent causation since it shows no sign of being "natural."

Apparent Design
Objects which contain features that most observers would acknowledge as giving the appearence of design.

Behe, Michael (1952 - Present)
Professor, biochemist and author of *Darwin's Black Box*, Behe is one of the best known advocates of Intelligent Design and, specifically, the concept of irreducible complexity. He is also a fellow at the Center for Science and Culture as well as a member of the Discovery Institute.

Choice
Mental process of thinking involved with judging the merits of multiple options and selecting one; directed contingency.

Class of Models
Sufficiently similar and worked-out hypotheses and scientifically testable implications so that they have the potential of making significant

common predictions and providing significant common explanations of phenomena.

Coevolution
A change in the genetic composition of one species (or group) in response to a genetic change in another.

Complexity
The condition of being multi-faceted, intricate, and compounded.

Conjectural Design
A phenomenon in which design is claimed to be a part of its causal history, yet the claim is not well-founded because of a lack of sufficient evidence.

Contextuality
A phenomenon exhibits counterflow based on it's location or context.

Counterfactual Substitution
The direct violation of a natural law.

Counterflow
Events or phenomena that result contrary to what, in the relevant sense, would (or might) have resulted or occurred had nature operated freely.

Darwinism – theories of evolution deriving from the work of Charles Darwin and Richard Wallace, as published from 1858 – 9 on. Subsequently, in the 1920's – 40's, in light of developments in genetics and related studies of evolutionary population dynamics a neo-darwinian synthesis led to the classical form of the modern evolutionary theory. Currently, this is undergoing changes in light of various observed and proposed mechanisms such as horizontal gene transfer and the like.

Darwin, Charles Robert (1809 - 1882)
British naturalist and author of the *Origin of Species* and numerous other books on biology, Darwin is generally credited with the establishment of evolution as a serious science and with originating the core concept of natural selection.

Dawkins, Richard (1941 - Present)
British ethologist (animal behaviorist), writer and atheist, Dawkins is often referred to as Darwin's Rottweiler for the vehemence of this arguments favoring evolution—much as Thomas Huxley was referred to as Darwin's Bulldog more than a century earlier. Dawkins is the author of numerous books and publications on evolution, including *The Selfish Gene* in 1976 and a more recent response to Intelligent Design entitled *The Blind Watchmaker: Why the Evidence of Evolution Reveals a Universe without Design*. His latest book is *The Greatest Show on Earth* 2009.

Dembski, William Albert (1960 - Present)
Mathematician, creationist and author of *Science and Evidence for Design in the Universe*, Dembski is one of the best known advocates of Intelligent Design and, specifically, the concept of specified complexity.

Deep Counterflow
Counterflow is exhibited through non-apparent involvement in the phenomena being studied.

Design
Purposeful arrangement of parts.

Design Actualization
The formation of parts and/or the assembly of component parts into a system that functions to accomplish the original purpose.

Design Analogues
Physical effects of intelligence which hold across the spectrum of all possible intelligences, including the effects of human intelligence and any other intelligence, given the constraints on physical activity

(forces, space, time, energy, chance, and their interaction) imposed by the universe.

Design Conceptualization
The act of mind in the action of designing something for the accomplishment of a purpose.

Design Product
The physical result(s) of design conceptualization and actualization.

Designoid
An object in nature that appears to be designed, but actually intelligent agency had no part in its causal history.

Designer
An intelligent agent that arranges material structures to accomplish a purpose. Whether this agent is personal or impersonal, conscious or unconscious, part of nature or beyond nature are live possibilities within the theory of intelligent design.

Diachronic Counterflow
Counterflow that is perceivable chronologically over time.

Einstein's Axiom
A hypothetical thought experiment employed by the theoretical physicist Albert Einstein when considering how research would be impacted by axiomatically assuming that the universe was designed by an intelligence.

Electional Adaptation
Apparently intentional adaptation of means to ends, which do not appear to be random, or even iterative. Instead the choice for high fitness is taken immediately, with this high-fitness choice occurring repeatedly in different elective events.

Electional Exaptation
Apparently intentional adjustment of ends to means, which do not appear to be random, or even iterative. Instead the choice for high fit-

ness is taken immediately, with this high-fitness choice occurring repeatedly in different elective events.

Eliminative Design
Detecting design by systematically eliminating all other possible causes.

Evolution
Change over time; history of nature; any sequence of events in nature The theory that species on earth arise from simpler forms due to natural selection. Further, the belief that all life on earth originated by scientifically explicable, natural processes, without any need for an external guiding agent.

Exogenous Seeding
Teleological - The concept that a heterogeneous consortium of unicellular organisms that were products of advanced bioengineering and were used to seed the earth with life.

Fuller's Hypothetical
Hypothetically viewing phenomena in the universe as designed, whether the researcher holds that the objects of study are actually designed or not, for the express purpose of deriving novel data, hypotheses, experiments, and practical applications.

Functional Design
A pattern embodying counterflow which also performs or provides a novel function.

Governance
The quality of a process for which design was a systemic cause of the purpose, arrangement, and operation.

Hard Counterflow
Counterflow is exhibited in such a way that the prerequisite knowledge to identifying the counterflow involved only demands a familiarity with natural regularity.

Incredulous Design
Awe that is inspired by certain features of reality.

Informational design
Purposeful arrangement of parts that yields informational properties, which have a variational quality being distinctly different from order and regularity on the one hand, and disorder and chaos on the other.

Innovation detection
Searching for the presence of design input within a preexisting aboriginal design.

Intelligent agency
The recognition and verification that a cognitive process was at work in the causal history of a phenomenon or event.

Intelligent Design
The science of studying the criteria, parameters and procedures for reliably detecting the activity of an intelligent agent.

Intelligibility
A phenomenon or event with a recognizable signal or pattern that can be understood and is comprehensible objectively or extrinsically; understandability.

Interlaced design
Condition of an object or event that has some combination of chance, natural regularity, and design as part of its causal history.

Irreducible Complexity, IC
A system performing a given basic function is irreducibly complex if it includes a set of well-matched, mutually interacting, nonarbitrarily individuated parts such that each part in the set is indispensable to maintaining the system's basic, and therefore original, function. The set of these indispensable parts is known as the irreducible core of the systement.

Isomorphic Design
Organisms which are morphologically and phenotypically identical in a manner which exhibits counterflow, while being genetically different.

Isomorphic Instantiation
Complex technology developed by intelligence is subsequently found to exist in nature.

Law of Accumulating Criteria
An inference to design becomes stronger as more design criteria are satisfied by a

Lurking Variable
A statistical variable that has an important effect on either the response variable(s) or predictor variable(s), or both,

Macroevolution
This somewhat ambiguous term generally refers to large evolutionary developments such as speciation. In the debate over evolution, many antievolutionists concede microevolution (small changes) as a legitimate scientific process while reserving macroevolution for God or an Intelligent Designer.

Metascientific
An hypothesis that serves to provide a relatively broad framework from within which science can be practiced in a certain domain, or multiple domains.

Meta-theory
A theory developed to analyze theoretical systems.

Microevolution
This somewhat ambiguous term generally refers to small evolutionary developments such as changes in skin or hair color within a species. In the debate on evolution, many antievolutionists concede microevolution as a scientific process while reserving macroevolution for God or an Intelligent Designer.

Natural Selection
This is Darwin's explanation of how natural organisms adapt to and survive in nature—the process by which the randomness of mutation becomes the order of evolution.

Mind Correlative
The fittingness of understanding that exists between mind and phenomenon, seemingly making knowledge and science possible.

Modular Design
The arrangement of parts in a phenomenon give it the ability to change form in a series of steps, whether as self-reproduction or growth.

Natural Regularity
The qualities of the universe governing the behaviors of matter and energy, typically seen as natural forces, constants, and laws.

Natural Selection
This is Darwin's explanation of how natural organisms adapt to and survive in nature—the process by which the randomness of mutation becomes the order of evolution.

Nature
The essential qualities by which the universe is recognized.

Necessary Design
Object or event for which the only causal explanation is design.

Nomic Discontinuity
Effects that do not follow from the normal operations of natural law which can serve as an indicator of design.

Non-telic Explanation
An accounting of events that has necessary and strict recourse to natural regularity.

Ontological Evidence
Data that would be expected if a hypothesis is true.

Origin of Life
This is the theory of how life *started*, as opposed to evolution which is the theory of how life *developed*.

Paley, William (1743 - 1805)
English philosopher Paley was one of the first advocates of a form of design from intelligence, and his 1803 Argument from Design is the intellectual predecessor of modern Intelligent Design. Darwin acknowledged Paley's views in *Origin*, and specifically addressed these views in his discussion of the vertebrate eye.

Rarefied Design
Phenomena which exhibit qualities that can be very reliably claimed as having intelligence involved in their causal history.

Retrodiction
Retrodiction is the act of making a prediction about what science may find while looking at evidence from the past.

Science
A systematic study undertaken to obtain knowledge about nature that can be empirically verified.

Serendipity
Making unexpected and fortunate discoveries, often when looking for something else.

Spandrel
In evolutionary biology, a spandrel is a phenotypic characteristic that evolved as a side effect of a true adaptation.

Statistical Probability
Detecting design by probabilistically eliminating all other possible causes.

Surface Counterflow
Counterflow is clearly exhibited through the phenomena being studied.

Synchronic counterflow
Counterflow is perceivable in an instantaneous fashion, e.g. a photograph of the phenomenon will exhibit its counterflow.

Systemic counterflow
Counterflow is exhibited through the entire process involved in the phenomena being studied.

Teleology
Teleology is the supposition that there is design, purpose, a preconceived goal, directive principle, or finality in the works and processes of nature.

Theoretical Kind
A posited place-holder, delineated by observed causal effects.

Tractability
The necessary nature of an hypothesis such that through scientific and empirical means we can develop and test models of its "internal dynamics"

Wells, Jonathan, PhD. A senior member of the Discovery Institute, and author of the counter evolution book, *Icons of Evolution*. He holds PhDs from Yale and the University of California, and has been published widely.

Courtesy of Research ID in part

About the Author

EDMUND T. DOMBROWSKI, M.D., INC.

Diplomate, American Board of Orthopaedic Surgery
Qualified Medical Examiner

PERSONAL RESUME
(Prepared October 2002)

PERSONAL BACKGROUND

Edmund T. Dombrowski is a 79-year-old, orthopaedic surgeon residing in Redlands, California (36 years). Dr. Dombrowski retired from the surgical practice of orthopaedics in October, 1997. He and his wife, Eleanor Lucille (Reagan), have five children. Dr. Dombrowski is a member of the Church Jesus Christ of Latter-day Saints. He is also an instrument pilot with multi-engine, commercial and sea plane ratings. He is also a Life Scout, a member of the "Order of the Arrow," and a former two-term president of the Grayback Council of the Boy Scouts of America.

EDUCATIONAL BACKGROUND

- B.A., 1952, UCLA (Freshman letterman in swimming and water polo - three year water polo letterman)
- M.D., 1956, UCLA
- Internship, 1957, University of Colorado
- Orthopaedic Fellowship, 1957-1961, Mayo Clinic

MILITARY SERVICE

- Two years of active duty, Lt. Commander, Great Lakes Naval Hospital, Great Lakes,
- Illinois
- Six years, United States Naval Reserve

PROFESSIONAL ORGANIZATIONS

- Diplomate, American Board of Orthopaedic Surgery.- Jan, 1965
- Founding Member, North American Spine Society
- Member, American Academy of Orthopaedic Surgeons - emeritus
- Member, western Orthopaedic Association -emeritus
- Member, American Medical Association
- Member, San Bernardino County Medical Association - emeritus
- Former Member, UGA -Flying Physicians
- Former Treasurer, American College of Spinal Surgeons
- Former Frater, American College of Surgeons
- Former Member, International Society for the Lumbar Spine (world-wide membership at the time limited to 150)

HOSPITAL AFFILIATIONS

- Redlands Community Hospital
- Assistant Clinical Professor, Orthopaedic, Loma Linda. University Medical Center - emeritus

CONTRIBUTIONS TO THE MEDICAL ARMAMENTARIUM

A. Papers presented to medical societies:

1. Chicago Orthopaedic Society, 1962 -Closed Irrigation Suction Treatment of Osteomyelitis
2. Armed Forces Orthopaedic Society, 1963 -Closed Irrigation Suction Treatment of Osteomyelitis
3. American Academy of Orthopaedic Surgeons, 1964 -Closed Irrigation Suction Treatment of Osteomyelitis
4. San Diego Orthopaedic Society, 1965 - Understanding Anatomy and Physiology of Osteomyelitis
5. Western Orthopaedic Association, 1967 -The Chronic Dural Compression Syndrome

6. UCLA Alumni Association, 1968 -Lumbar Root Compression Due to Chronic Dural Compression
7. Loma Linda Alumni Association; 1968 - Lumbar Root Compression Due to Chronic Dural Compression
8. Western Orthopaedic Association, 1974, Progressive Spinal Stenosis
9. International Society for-the study of the Lumbar Spine, 1975 - A New Approach to Surgery of the Lumbar Spine
10. Loma Linda University, :1975, 1976, 1978, 1979, 1980, 1981, 1982 –Surgical Anatomy of the Lower Extremities
11. Numerous community and political organizations, 1978 through 1981 -The True Cause of Increasing Medical Costs
12. Numerous community and political organizations, 1978 through 1981 -The Effect of an Aging Population on a Society's Standard of Living
13. Los Angeles Chapter, Western Orthopaedic Association, 1982 -Long-term Results of Patellectomy
14. Annual Meeting of the Western Orthopaedic Association, Albuquerque, New Mexico, October 1982 -Patellectomy: An Excellent and Definitive Procedure
15. Leeds, England, January, 1982 - A New Approach to the Management of Spinal Injury
16. Los Angeles Chapter, Western Orthopaedic Association, SantaBarbara, California, May, 1983 -Spinal Fixator for Spinal Injuries
17. Annual meeting of Orthopaedic Association, Salt Lake City, Utah, October, 1983 - Comparison of Pantopaque and Metrizamide Myelography
18. The combined meeting of Orthopaedic Surgeons and Neurosurgeons, Las Vegas, Nevada, October, 1983 - Results of Fat Transplanted into the Epidural Space
19. The First International Congress on Advances in Orthopaedics and Traumatology, Palm Springs, April, 1984 - The Rezaian Spinal Fixator for Spinal Fixation

20. The First International Congress on Advances in Orthopaedics and Traumatology, Palm Springs, April, 1984 - Wide Decompressive Laminectomy and Fusion for Lumbar Root Compression

21. The first Internal Congress on Advances in Orthopaedics and Traumatology, Palm Springs, April, 1984 - Diagnosis and Treatment of Acquired Spinal Stenosis

22. The first Annual Meeting of the American College of Spinal Surgeons in conjunction with the annual meeting of the American Academy of Orthopaedic Surgeons, Las Vegas, Nevada, February, 1985 - Wide Decompressive Laminectomy and Fusion for Lumbar Root Compression

23. The Western Spinal Clinic Workshop, Salt Lake City, Utah, March, 1985 - Evoked Cortical Potentials in the Diagnosis and Treatment of Lumbar Root Compression

24. The Western Spinal Clinic Workshop, Salt Lake City, Utah, March, 1985 - Diagnosis and Treatment of the Failed Back Syndrome

25. The Western Spinal Clinic Workshop, Salt Lake City, Utah, March, 1985 - Indications for Lumbar Fusion

26. The Western Spinal Clinic Workshop, Salt Lake City, Utah, March, 1985 - Technique and Results of Lumbar Transverse Process Fusion

27. The Western Spinal Clinic Workshop, Salt Lake City, Utah, March, 1985 - Diagnosis and Treatment of Spinal Stenosis

28. National Sub-Acute Association, Washington D.C., August, 1988 – Matching Severity of Disease to Intensity of Care

29. National Association of Outpatient Surgeons, Newport Beach, California, June, 1990 – "Sub-Acute Care"

30. AME/QME Seminars, Los Angeles, California, February, 1992 - Disability and the Orthopaedic Examination

31. NME Seminar, Ontario, California, December 10, 1993 - Managed Care in Spine Surgery

32. Chino Medical Center Staff Seminar - The California Workers' Compensation System - July, 1995
33. American Academy of Orthopaedic and Neurologic Surgeons - Las Vegas, Nevada, August, 1995: Unilateral Steffee Plating, 61 Cases, Single and Multiple Levels of the Lumbar Spine
34. Association of Orthopaedic and Neurologic Surgeons - Las Vegas, Nevada - August, 1995: The Lethal Cell Saver Syndrome
35. 8th International Conference on Lumbar Fusion and Stabilization -San Antonio, Texas - September 30 through October 4, 1997 - 61 Cases of Unilateral Fixation, Single and Multiple Levels
36. Annual Meeting Physicians Premier Services Corporation - Los Angeles and San Francisco: Introduction to Workers' Compensation

B. Exhibits:

1. American Academy of Orthopaedic Surgeons, 1964, The Treatment of Osteomyelitis by Closed Irrigation Suction
2. American College of Surgeons, 1964, The Treatment of Osteomyelitis by Debridement and Closed Irrigation Suction
3. North Carolina Medical Association, 1965, The Treatment of Osteomyelitis by Debridement and Closed Irrigation Suction
4. Lorna Linda Alumni Association, 1965, The Chronic Dural Compression Syndrome
5. American Academy of Orthopaedic Surgeons, 1969, The Diagnosis and Treatment of the Chronic Dural Compression Syndrome – Secondary to Spinal Stenosis
6. American Medical Association, 1969, The Diagnosis and Treatment of the Chronic Dural Compression Syndrome – Secondary to Spinal Stenosis
7. Loma Linda Alumni Association; 1983 (three separate exhibits concerning Patellectomy, Leg Strengthening and Anterior Column Fixation for Spinal Injuries with Spinal Fixation

8. First International Congress for Advances in Orthopaedics and Traumatology, Pa1m Springs, 1984, A New Technique for Patellectomy, A New Technique for Leg Strengthening (with S.M. Rezaian), Spinal Fixator for Spinal Surgery (with S.M. Rezaian)
9. Hands Across the Sea Joint Meeting of the Japanese, Canadian and American Spine Societies, Maui, Hawaii, July 1994, The Lethal Cell Saver Syndrome

C. Former Yearly Lectures:

1. Bilroth Post-Graduate Lectures on the Anatomy of the Lower Extremities, Loma Linda University Medical Center, 1979-1984
2. Introduction to Orthopaedic Surgery, Loma Linda University Medical Center, 1970 through 1978
3. Congenital Anomalies of the Lower Extremities, Loma Linda University, 1978 through 1983
4. The Medical-Legal Orthopaedic Evaluation - Preparation Course for QME Certification, 1994
5. The Orthopaedic Examination in Workers' Compensation AME-QME Seminars

D. Published Papers:

1. The Closed Irrigation Suction Treatment of Osteomyelitis - Orthopaedics
2. The Chronic Dural Compression Syndrome - Journal of Bone and Joint Surgery
3. Progressive Spinal Stenosis - Journal of Bone and Joint Surgery
4. Patellectomy: An Excellent and Definitive Procedure, Abstract - Journal of Bone and Joint Surgery
5. Patellectomy: An Excellent and Defuiitive Procedure, Abstract – Orthopaedics Today

6. Spinal Fixator for the Management of Spinal Injury (The Mechanical Rationale) MEP, Ltd, 1983, Vol. 12
7. Spinal Fixator for Surgical Treatment of Spinal Injury - Orthopaedic Review
8. Efficacy of Morbidity of Water Soluble and Oil-Based Myelography - Orthopaedic Review, Vol. 15, No.1
9. Rezaian Fixator in the Stabilization of the Unstable Spine - Orthopaedic Review, Vol. 15, No. 1, January, 1986
10. Requested for Publication by the Journal of Orthopaedic and Neurologic Surgeons: 61 Cases of Unilateral Internal Fixation of the Lumbar spine

CONTRIBUTIONS TO THE BUSINESS AND PROFESSIONAL COMMUNITY:

1. Multiple lectures on methods of health care cost containment.
2. Understanding the new California HCO legislation, April 30, 1994, San Francisco, California.
3. Understanding the new California HCO legislation, May 1, 1994, Los Angeles, California.
4. The value of the new California HCO legislation to the business community, Ontario, California, October, 1994.
5. Understanding the value of California's HCO legislation to the employer, Pomona, California, February, 1995.
6. Understanding the value of California's HCO legislation to the employer, Loma Linda, March, 1995.

SPECIAL MEDICAL HONORS:

- Certificate of merit, American Medical Association (exhibit on spinal stenosis, 1969)
- Former Assistant Clinical Professor of Orthopaedic Surgery, LLUMC - retired
- Former Assistant Clinical Professor of Orthopaedic Surgery, COMP

- Former Member, Federal Council on Aging, 1983 - 1985 - Counselors to the President of the United States - regarding legislation affecting persons over 65 years of age
- Former Ex-Officio Member of the National Advisory Council on Aging (NACA), 1983 tc 1985 -Allots NIH grants in the field of aging
- Former Member, 12th District Board of Medical Quality Assurance, Review Committee, State of California, 1979-1982
- Chairman of the first session of the first meeting of the American College of Spinal Surgeons (amalgamated with North American. Spine Association to Form North American Spine Society)
- Former Member, Council of Musculoskeletal Specialty Societies of the American Academy of Orthopaedic Surgeons
- Nominated to be a member of the Spine Committee of the American Academy of Orthopaedic Surgeons - declined
- Former Member, Credentials Committee, Western Orthopaedic Association (Los Angeles Chapter)
- Former Orthopaedic Consultant to CMRI
- Qualified Medical Examiner, State of California, 1994

RECOGNITION FOR DISTINGUISHED ACHIEVEMENTS:

- Past National Chairman, Committee for a Week of National Unity, 1969 (Bob Hope, Honorary Chairman)
- Past President, Kimberly Crest Association, Inc. - A project which converted an old mansion in Redlands into a permanent museum representing the late 1800's and preserving mementos of Redlands history
- Past President, Friends of Prospect Park - Succeeded in buying a $420,000, 40-acre arboretum for the City of Redlands through private subscription (current value more than $4,000,000)
- Past President and Charter Member, Redlands - Yucaipa Guidance Clinic – An organization designed to improve men-

tal health in the Redlands area by helping underprivileged children and by helping teenagers who are engaged in drug abuse

- Past Chairman, 1968 4th of July parade. Reinstituted the 4th of July celebration in
- Redlands after 30 years of nonactivity
- Grand Marshal, 1969 4th of July parade
- Past Member, Board of Directors - World Affairs Council of Redlands-Riverside-San Bernardino
- Recipient of Elks and Chamber of Commerce "Man of the Year" award, 1968
- Member, Kimberly Crest Association and Friends of Prospect Park, 1969-1982
- Member, Board of Directors, Kimberly Crest Association, 1969-1982
- Chairman,. LDS Inland Concert Association, 1982
- Former Member, Public Service Committee, San Bernardino Medical Association, 1982-1983
- Former National Council Member (35th Congressional District), National Center for Constitutional Studies
- Former President, Western Orthopaedic Institute - a scientific research organization
- Former Member, Medical-Legal Committee, San Bernardino County Medical Association
- Listed in Marquis, "Who's Who in the West"
- Listed in Marquis, "Who's Who in California"

BUSINESS EXPERIENCE

- Former President, Triple "D" Orchid Distributors
- Former President, E&E Computer Leasing
- Former Managing Partner, Redlands Hospital Hotel
- Former Managing Partner, C&D Aircraft Leasing

- Former Member, Board of Directors (and Secretary/Treasurer, Doctor's Hospital of Redlands)
- Rental Properties, Redlands, California and Star Valley, Wyoming
- Former President and CEO, General Healthcare Corporation (applicant for the WCHPO segment of California's HCO legislation under the name of HCO Management Corporation)
- President Physicians Premier Services Corporation

PATENTS

- Stabilizing clip for spinal Instrumentation - Granted 1994
- Anti-pull out plug -pending

Made in the USA
Monee, IL
25 October 2026

66550177R00098